T0324812

Synthetic Biology

Synthetic Biology

A Sociology of Changing Practices

Andrew S. Balmer
University of Manchester, UK

Katie Bulpin
University of Manchester, UK

and

Susan Molyneux-Hodgson
University of Sheffield, UK

palgrave
macmillan

First published 2016 by
PALGRAVE MACMILLAN

The authors have asserted their rights to be identified as the authors of this work
in accordance with the Copyright, Designs and Patents Act 1988.

Palgrave Macmillan in the UK is an imprint of Macmillan Publishers Limited,
registered in England, company number 785998, of Houndmills, Basingstoke,
Hampshire, RG21 6XS.

Palgrave Macmillan in the US is a division of Nature America, Inc.,
One New York Plaza, Suite 4500 New York, NY 10004–1562.

Palgrave Macmillan is the global academic imprint of the above companies
and has companies and representatives throughout the world.

Hardback ISBN: 9781137495419
E-PUB ISBN: 9781137495433
E-PDF ISBN: 9781137495426
DOI: 10.1057/9781137495426

Distribution in the UK, Europe and the rest of the world is by Palgrave
Macmillan®, a division of Macmillan Publishers Limited, registered in England,
company number 785998, of Houndmills, Basingstoke, Hampshire RG21 6XS.

Library of Congress Cataloging-in-Publication Data is available from the Library
of Congress

A catalog record for this book is available from the Library of Congress

Library of Congress Cataloging-in-Publication Data

Balmer, Andrew S., 1983–
    Synthetic biology : a sociology of changing practices / Andrew S. Balmer,
University of Manchester, UK, Katie Bulpin, University of Sheffield, UK, Susan
Molyneux-Hodgson, University of Sheffield, UK.
        pages cm
    Includes bibliographical references and index.
    ISBN 978–1–137–49541–9 (hardback)
        1. Synthetic biology. 2. Bioengineering. I. Bulpin, Katie, 1982–
II. Molyneux-Hodgson, Susan. III. Title.

TA164.B35 2016
660.6—dc23   2015033224

Contents

Acknowledgements

First, we must thank our colleagues and collaborators with whom we worked during our investigations into synthetic biology, from several science and engineering departments across a number of UK universities. Special mentions are due to Catherine, Phil, Graham, Jags, Joss, Greg, Joby, Simon and Qaiser. Thanks also to the iGEM team, who were an inspiration and a lot of fun: Caroline, Matt, Narmada, Nii, Steve and Tom. We have shared in prosperity and pleasure, failure and frustration. There was even a dinosaur at one point. We hope that you all enjoy the book.

Second, thanks to the creative and thoughtful community of sociologists involved in the ESRC Seminar Series on Synthetic Biology and the Social Sciences, especially Jane Calvert, Claire Marris and Emma Frow. We are grateful to our colleagues, postdocs and PhD students for helping us to thrash out some of the ideas in the book, and for their support more generally. Of course, thanks also go to the funders that helped to make this book possible: the EPSRC (EP/H023488/1), BBSRC (BB/F018681/1, BB/M017702/1), ESRC and the White Rose Science Education Network.

In a book about everyday life and knowledge production it is most important to show our appreciation to our friends, families and partners. You have all been so kind and patient. Andy is particularly grateful to Jamie, Mike, Ian and Suzie, Diane and Ivy for their support. Susie would like to thank Pete, Rosie (yes, you are more important than a book) and Alex. Kate thanks all the Bulpins and Morans, but especially Paul.

Finally, we each want to thank each other, just so that it is on the record. We had fun.

List of Abbreviations

AHRC	Arts and Humanities Research Council (UK)
BBSRC	Biotechnology and Biological Sciences Research Council (UK)
CCW	Consumer Council for Water
CSynBI	Centre for Synthetic Biology and Innovation
DBIS	Department for Business, Innovation and Skills (UK)
DWI	Drinking Water Inspectorate (UK)
EA	Environment Agency (UK)
ELSI	ethical, legal and social implications
EPSRC	Engineering and Physical Sciences Research Council (UK)
ESRC	Economic and Social Research Council (UK)
FOG	fats, oils and greases
GMOs	genetically modified organisms
HCSTC	House of Commons Science and Technology Committee
iGEM	International Genetically Engineered Machine Competition
JCVI	J. Craig Venter Institute (USA)
MIT	Massachussetts Institute of Technology (USA)
NIBB	Networks in Industrial Biotechnology and Bioenergy
NSF	National Science Foundation (USA)
Ofwat	Water Services Regulation Authority (UK)
R&D	Research and Development
RAE	Royal Academy of Engineering (UK)
RCUK	Research Councils United Kingdom
Roadmap	'A Synthetic Biology Roadmap for the UK' Report by the SBRCG
RRI	responsible research and innovation
SBRC	synthetic biology research centre
SBRCG	Synthetic Biology Roadmap Coordination Group (UK)
SBX.0	International Synthetic Biology Conference
SIM	service incentive mechanism
SSK	sociology of scientific knowledge
STS	science and technology studies
SYNBERC	Synthetic Biology Engineering Research Center (USA)

SynbiCITE   Synthetic Biology Innovation and Commercialisation
            Industrial Translation Engine
WFD   Water Framework Directive (EU)
WWICS   Woodrow Wilson International Center for Scholars

1

Synthetic Biology in Situ

> There is nothing more difficult to handle, more doubtful of success, and more dangerous to carry through than initiating changes. (Machiavelli, 1961: 21)

What is synthetic biology? Eight artefacts from the field

1. Quotation from an interview with a project participant.

Andy Balmer: You say you're working on a synthetic biology project. Why do you call it that?

Academic Molecular Biologist 1: Why do I call it that? Partly because that is where the idea [for my work] came from. The network I got involved with was badged as a synthetic biology network and the idea came out of that. But it may have equally come out of a network that wasn't badged as synthetic biology. It would have just been called biotechnology or something in the past. Would we have called it synthetic biology five years ago? Probably not. We would have just said it was a kind of microbial biotechnology approach or something.

2. UK Secretary of State for Business Innovation and Skills, the Rt Hon. Dr Vince Cable MP, speaking at the Manchester Institute of Biotechnology on 29 January 2015, following the launch of three synthetic biology research centres (SBRCs) funded by the Biotechnology and Biological Sciences Research Council (BBSRC).

We are investing in a series of centres to develop the science of synthetic biology, which is, in crude terms, using bugs to produce

1

chemicals rather than using hydrocarbons. It is an area where Britain, and Manchester particularly, excel. ... We believe in science and technology. The Government has been under a lot of financial pressure from the beginning – the deficit and all that. But we do believe that the long-term future of the country is as a knowledge economy. We have to invest in it with government money. The private sector's not going to do the very long-term, high-risk research. Universities can do that, together with the innovation centres we've started around the country.

3. Figure 1.1 A student in the International Genetically Engineered Machine Competition (iGEM) extracts standardised sequences of DNA, known as 'BioBricks™', from a 384-well plate. He will then pipette them out for amplification via a polymerase chain reaction, before transforming them into *E. coli* bacteria

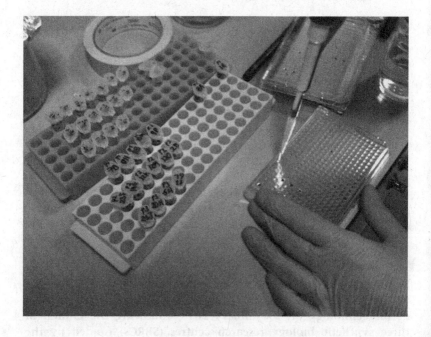

4. A quotation from 'Synthetic Biology – The State of Play', a 2012 review article by Prof. Richard Kitney and Prof. Paul Freemont, Co-Directors of the Engineering and Physical Sciences Research

Council (EPSRC) UK National Centre for Synthetic Biology and Innovation (CSynBI).

There is, in some quarters, still doubt about the definition of synthetic biology. This is not a view held by the international synthetic biology community. (The community can be defined as people who attend the major international "SB X.0" conferences and regularly organise teams for the International Genetically Engineered Machine Competition (iGEM) – a prestigious student competition involving many of the world's leading universities.) The accepted definition is 'synthetic biology aims to design and engineer biologically based parts, novel devices and systems – as well as redesigning existing, natural biological systems.' Synthetic Biology is the application of systematic design – using engineering principles. (Kitney and Freemont, 2012: 2029)

5 'To live, to err, to fall, to triumph, to recreate life out of life.' A quotation from James Joyce's novel, *A Portrait of the Artist as a Young Man*, which the J. Craig Venter Institute encrypted as a 'watermark' in the DNA of their 'synthetic' living microorganism, *Mycoplasma laboratorium*.

```
CAACTGGCAGCATAAAACATATAGAACTACCTGCTATAAGTGATA
CAACTGTTTTCATAGTAAAACATACAACGTTGCTGATAGTACTCCT
AAGTGATAGCTTAGTGCGTTTAGCATATATTGTAGGCTTCATAATA
AGTGATATTTTAGCTACGTAACTAAATAAACTAGCTATGACTGTAC
TCCTAAGTGATATTTTCATCCTTTGCAATACAATAACTACTACATC
AATAGTGCGTGATATGCCTGTGCTAGATATAGAACACATAACTAC
GTTTGCTGTTTTCAGTGATATGCTAGTTTCATCTATAGATATAGGC
TGCTTAGATTCCCTACTAGCTATTTCTGTAGGTGATATACGTCCAT
TGCATAAGTTAATGCATTTAACTAGCTGTGATACTATAGCATCCCC
ATTCCTAGTGCATATTTTCATCCTAGTGCTACGTGATATAATTGTA
CTAATGCCTGTAGATAATTTAATGCCTGGCTCGTTTGTAGGTGAT
AATTTAGTGCCTGTAAAACATATACCTGAGTGCTCGTTGCGTGAT
AGTTCGTTCATGCATATACAACTAGGCTGCTGTGATATGGTCACT
GCCCTTACTGTGCTACATATTACTGCGAGGGGGATGACGTATAA
ACCTGTTGTAAGTGATATGACGTATATAACTACTAGTGATATGACG
TATAGGCTAGAACAACGTGATATGACGTATATGACTACTGTCCCA
AACATCAGTGATATGACGTATACTATAATTTCTATAATAGTGATAAA
TAAACCTGGGCTAAATACGTTCCTGAATACGTGGCATAAACCTG
GGCTAACGAGGAATACCCATAGTTTAGCAATAAGCTATAGTTCGT
CATTTTTAA
```

6. Figure 1.2 A sticky note written by the authors in planning the proposal for this monograph. In the note we state that there are such things as 'actual synthetic biology projects'.

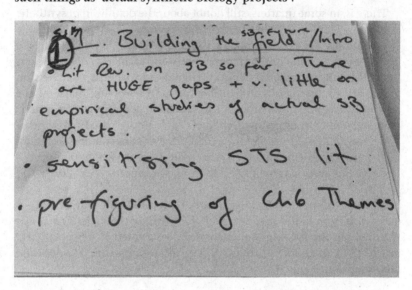

7. Lionel Clarke, Co-Chairman of the UK Synthetic Biology Leadership Council speaking about the 'Synthetic Biology Roadmap for the UK'.

> Synthetic Biology may be defined not only in terms of *what it is*, but also in terms of *what benefits it can deliver*: It has the potential to deliver important new applications and improve existing industrial processes – resulting in economic growth and job creation. (Clarke, 2013: italics in the original)

8. An equation created by an iGEM team as part of its endeavour to genetically engineer a strain of *E. coli* to act as a biosensor for cholera. The equations try to work out how much cholera needs to be present to turn the sensor on and off.

$$\frac{dL1p}{dt} = Input * k1(L1tot - L1p) - k2 * L1p * (L2tot - L2p)$$

$$\frac{dL2p}{dt} = k2 * L1p * (L2tot - L2p) - k3 * L2p * (L3tot - L3p)$$

$$\frac{dL3p}{dt} = k3 * L2p * k1(L3tot - L3p) - k4 * L3p * (L4tot - L4p)$$

$$\frac{dL4p}{dt} = k4 * L3p * (L4tot - L4p) - k5 * L4p$$

What is synthetic biology? A review of the literature

Synthetic biology is in the making. At the turn of the second millennium, academic literature heralded the arrival of a new bioengineering field, promising to deliver a suite of technical fixes for seemingly every global problem. Across the world, it seems that academics, industrialists, politicians, regulators, campaigners and a variety of other professions and publics have been caught up in this latest technoscience. For example, not only does it promise to revolutionise the fuel production system, but also to help to stabilise and reverse global climate change, all whilst minimising political and economic energy volatility (Savage *et al.*, 2008) And that is not all it has promised the world. From the medical to the military, synthetic biology is positioned to provide solutions to society's most obstinate problems (McDaniel and Weiss, 2005).

The field is emerging onto a world stage on which global social challenges are increasingly prominent. It is now commonplace to frame technical research in terms of such challenges, whether these are curing diseases or providing jobs and economic growth. Alongside meeting our needs to be healed, heated and fed (Osborne, 2012), synthetic biology is recruited to the enterprise of nation building: for example, government investment 'will ensure the UK remains at the *forefront* of synthetic biology' and 'the UK can be *world-leading* in this emerging technology' (DBIS, 2013: italics added for emphasis). Science and innovation have become increasingly visible as a territory on which claims to competitiveness can be made (Slaughter and Rhoades, 1996). As such, technical communities promising novelty and international competitiveness are rewarded by governments seeking an edge.

A prominent researcher in synthetic biology says:

> In our research, we're mixing different genes from different organisms in order to do new chemistry inside living cells ... The goal is to enable us to produce new drugs for fighting disease or combating bioterror agents, or to produce existing drugs in better ways ... to harness the power of biology to solve problems that cannot be solved in any other way. (J Keasling, quoted in Yarris, 2004)

It is not unusual to find scientific discourse mobilising the future in this manner. Synthetic biology is only the latest in a series of 'promissory sciences' (Brown and Michael, 2003). Indeed, the field has gained buzzword status (Hedgecoe, 2003), such that the term 'synthetic biology' itself helps to construct a community (Molyneux-Hodgson and Meyer, 2009; Powell *et al.*, 2007). In a similar vein, actors in the field regularly make use of a new set of terms and metaphors, derived from engineering, to conceptualise the objects with which they work (Balmer and Herreman, 2009). Hellsten and Nerlich (2011) point out that there is some continuity with past biotechnology endeavours and the human genome project as regards the use of engineering metaphors. However, in synthetic biology, they argue that:

> The focus shifts from deciphering the blueprint of life to building life according to scientists' own blueprint, brick by BioBrick. Whereas in the past the metaphor "building blocks" of life was just a metaphor, it is becoming increasingly real and whereas in the past these blocks were, in a sense, made and put together by "nature," they are now designed and assembled (and controlled) by humans. (Hellsten and Nerlich, 2011: 394)

We return to this shift in the terminology from genes to 'parts' or BioBricks™ below. Of course, it is not only in promissory narratives and terminological shifts that the field is being made. Practices are also being reconfigured. In the early 2000s, a number of insider commentators on bioscience were quick to try to categorise the emerging practices being termed 'synthetic biology' in specific ways. As Benner and Sismour (2005: 533) claimed:

> Synthetic biologists come in two broad classes. One uses unnatural molecules to reproduce emergent behaviours from natural biology, with the goal of creating artificial life. The other seeks interchangeable parts from natural biology to assemble into systems that function unnaturally.

Social scientists were hot on their trail, and were similarly interested in what exactly synthetic biology might be. O'Malley *et al.* (2008: 57) concluded that:

> Synthetic biology...can be understood as encompassing three broad approaches towards the synthesis of living systems: DNA-based

device construction, genome-driven cell engineering and protocell creation.

In this way, a number of new or re-purposed practices were being identi-fied and then grouped in different ways by different observers during this early period of reflection on the usage of the term. Nonetheless, natural scientists, engineers and social scientists alike tended to see in these diverse practices an emphasis on the adoption of engineering and computational epistemology to take better control of the biolog-ical realm. In this regard, synthetic biology is one of the most recent incarnations of a lineage of engineering-based approaches to biology that have recurred throughout the twentieth century (Campos, 2010; Campos, 2012). However, it does seem that there is a distinctly utili-tarian 'drive to make' in the latest manifestation that animates the field as a whole (Schyfter, 2013a) and, as such, that synthetic biology represents a shift in the exact ways in which engineering and molecular biology are entangled.

One of the most prominent areas of research in which an engineering approach is being used is in so-called 'parts-based' synthetic biology. This work tends to invoke a set of design principles to pursue the modularisation and standardisation of biological materials into distinct 'parts', 'circuits', 'devices' and 'chassis' (Endy, 2005). The construction and use of standard-ised parts with characterised properties and predictable functions is under-stood to facilitate the 'rational' design and assembly of complex systems with equally robust and predictable behaviours (Kitney and Freemont, 2012). The parts-based approach is tied to the development of standard-ised cloning technologies and the refinement and decreasing cost of DNA synthesis (Shetty *et al.*, 2008). Whereas long-standing cloning methods have tended to rely on 'parts' already existing in nature, the advent of DNA synthesis techniques has meant that genetic sequences can now be designed *de novo* using computers and then chemically synthesised for assembly and incorporation into 'host' organisms, often called 'chassis'. The construction of these biological objects is currently conducted using a mix of traditional and novel molecular biology methods, but the ambition is to move towards more automated processes for assembly. An impor-tant shift in the practices for managing such parts has also occurred, as those working in the parts-based approach are encouraged to submit their standardised sequences into one of a number of Registries of Standard Biological Parts for others to use freely.

Sociological and philosophical studies of the parts-based approach have concentrated on unpacking the epistemological and methodological principles that underpin it (Keller, 2009; O'Malley *et al.*, 2008), as well as on the ontological implications for the kind of objects being imagined and produced (Holm and Powell, 2013; Schyfter, 2012). In particular, there has been considerable interest in how synthetic biologists are grappling with the ways in which the microbiological context affects the functional properties of standardised parts (Bennett, 2010a; Güttinger, 2013). In other words, how does the bacterial and biological environment within which DNA parts are situated affect their function? For engineers, the fact that a standard screw thread will work in a standard socket presents the mechanical equivalent of the biological goal to which they aspire (Knight, 2003). They wish to produce the same kinds of standardisation for biological parts, however the interaction between a biological 'chassis' and its biological 'content' creates a challenge that does not exist in the case of screw threads and sockets.

Seeking to do away with this 'problem of context' as best as they can, other researchers calling themselves synthetic biologists have developed synthesis technologies to design and manufacture entire genomes from scratch. John Craig Venter's eponymous Institute (JCVI) is one of the largest companies conducting this kind of research. A prominent example of their approach is the creation and transplantation of a synthetic 'minimal genome' into a bacterium to create a viable 'synthetic cell' (Gibson *et al.*, 2010). A similarly minimalist approach to synthetic biology can be found in the work of scientists and engineers concerned with manipulating and synthesising the membrane and lipid components of cells to create simple 'protocells' (Walde, 2010). Such efforts to minimise the complexity of the cell are generally hoped to increase understanding of the origins of life, particularly in the context of protocell development (O'Malley *et al.*, 2008), whereas in the creation of minimal genomes efforts are primarily geared towards taking greater control of the cell's activity and so also the predictability of its functions. Scientists and engineers like Venter, working in the area of minimal cells, hope that by creating organisms from scratch the process of industrialising biology will more neatly follow the 'design, build, test, improve' cycles found in other engineering fields, such as in the civic construction of buildings, bridges and so on.

Changes are anticipated in such research and innovation practices not only for the epistemological and ontological dimensions of biological objects but also for the identities and everyday lives of engineers and their students. Reshma Shetty, a prominent industrial synthetic biologist and co-founder of the biotech start-up Ginkgo Bioworks, describes how their facilities allow the synthetic biologist to 'spend more time thinking about the design, rather than doing the grunt work of making DNA' (Quoted in Singer, 2009). Entrepreneurial synthetic biologist Robert Carlson (2010: 124) anticipates a future for the field in which:

> most students and novice gene-hackers will be better at conceptualising and modelling new genetic circuits than building them. Where design experience will exceed practical experience, commercially available kits include simple recipes that allow moving genes between organisms.

In this regard, shifting practices of engineering microorganisms are organised in relation to changing how academics spend their time, train their students and experience the world. The implications of 'grunt work' in Shetty's remarks are telling. Synthetic biologists are often keen to point out how established biotechnology practices require a significant amount of embodied skill in the laboratory and that this comes at the expense of spending more time on design. It follows that computer-assisted design has become a central feature of how interdisciplinary relationships between engineers, computer scientists and systems biologists are envisioned and increasingly organised in the field. Mackenzie (2010: 181) argues that 'biological work, techniques and materials are being reconfigured under the rubric of design in synthetic biology.' A key material and infrastructural implication of this emphasis on shifting everyday scientific practices from the laboratory to the computer interface is that such mundane work with microorganisms has to be disembodied through the use of robotics. High-throughput screening, synthesis and testing technologies are understood to free up the labour time of human actors, so that standardisation and automation come with a correlative change in the kinds of skills required to be a synthetic biologist. Nonetheless, exactly how this reorganisation of labour is accomplished or imagined varies across different synthetic biology research groups.

As such, there remain diverse ways in which researchers are working with engineering approaches to manipulate and manufacture microorganisms. Underlying the use of engineering in these heterogeneous contexts, however, is the promise that having better control of living systems, through whichever means, will enable synthetic biology to have extraordinary industrial impacts. Indeed, it is not uncommon to hear the claim that the field will bring about a second industrial revolution, as engineering is understood to have created the first. As James Chappell states: 'One of the major goals in synthetic biology is to find a way to industrialise our processes so that we can mass produce these biological factories much in the same way that industries such as car manufacturers mass produce vehicles in a factory line' (quoted in Smith, 2013a). This analogy is particularly pertinent in the context of standardisation in the parts-based community, as Tom Knight espouses:

> William Sellers... argued forcefully and successfully for the standardization of pitch, diameter, and form of screw threads, providing the infrastructure which allowed the industrial revolution to take off. The machinists of the Franklin Institute built the lathes, drills, taps, and dies necessary to make the standard a success. We anticipate advantages similar to those which accompany the standardization of screw threads in mechanical design – the widespread ability to interchange parts, to assemble sub-components, to outsource assembly to others, and to rely extensively on previously manufactured components. (Knight, 2003: 2)

On this note, Mackenzie (2013: 74) points out that synthetic biologists thus seek to 'emulate the eminent successes of the civil engineering, software engineering and microelectronics industry.' Various corporate entities have been set up to try to realise this dream of a biological revolution, and major industries as well as venture capitalists have begun investing. This mobilisation of infrastructure is, however, dependent on productive relations between academia and business entities (Tait, 2010; Tait, 2012).

Government actors around the world have taken an interest in this emerging technoscience. Between 2005 and 2010 the United States government is reported to have spent around $430m on synthetic biology research (WWICS, 2010), much of which has come from the Department of Energy. In large part, this is because synthetic biology has promised to construct organisms capable of churning out industrial-scale quantities of biofuels at economically viable prices. The Synthetic Biology Engineering Research Center (Synberc), a multi-institutional US programme, funded to the tune of $16m by the National Science Foundation (NSF), has been a key player in constructing this vision and has made biofuels one of its primary targets for development.

In the UK, the coalition government (2010–15) played a major role in the constitution of synthetic biology as a key area of development for the UK Research Councils (RCUK), and so also for scientists and engineers, industrialists and regulators. In 2007, David Uffindel, the Head of the Bio-Economy Unit in the Coalition's Department of Business, Innovation and Skills (DBIS) convened a round-table meeting, under the instruction of the Rt Hon. David Willetts MP (Minister of State for Universities and Science) and the Rt Hon. Dr Vince Cable MP (Secretary of State for Business, Innovation and Skills). Synthetic biologists, industrialists, government actors and social scientists were brought together to discuss what role the government could play in helping to ensure the maximum economic impact of the emerging field. It was concluded – though there was a sense that this was a fait accompli – that a Synthetic Biology Roadmap Coordination Group should be established to create a roadmap for the development of the field in the UK. The Synthetic Biology Roadmap (SBRCG, 2012) has gone on to shape RCUK investment in the field, for example, in helping to establish six synthetic biology research centres across the country, costing around £75m. The UK funding councils, the Roadmap and the research centres are explicitly orienting synthetic biology to industrial goals, in order to accelerate the 'route to market' for university research. These moves can be understood as the material accomplishment of promises, in some regards. In our study we are concerned with elaborating how such promises are made sense of and incorporated, or not, into changes in practices, and the reconfiguration of material relations at the level of research projects through everyday academic and industrial life.

Observing the institutionalisation and industrialisation of synthetic biology, social scientists have shown how classifications like 'synthetic' and 'natural' are being constituted in relation to the ways in which epistemic ideals and commercial imperatives are negotiated within the bioscience community (Calvert, 2008; Calvert, 2010). The different ways in which the field is being enacted in relation to engineering epistemology, material infrastructure, sharing parts and commercial imperatives are all implicated in the constitution of quite a messy intellectual property landscape. It is still unclear whether synthetic biology will follow the precedents set by gene patenting or turn into what Calvert (2012: 169) calls a 'diverse ecology of the open and the proprietary.' This will probably be determined by how industrial and governance actors shape the emergence of the field in the coming years.

What is more certain is that the economics and ethics of synthetic biology are tied into emerging practices of research and community formation. A core strand of sociological work has been concerned with this issue, examining how new communities of academic and industrial actors are being assembled (Molyneux-Hodgson and Balmer, 2014; Schyfter *et al.*, 2013). Sociologists working in science and technology studies (STS) have identified specific normative frameworks around which these new communities are coalescing. This can be seen, for example, in the parts-based approach where the intertwined objectives of community building, standardisation and the creation of an 'open source' sharing ethos continue to be debated (Calvert, 2012; Frow and Calvert, 2013a). This community appears to be crystallising around a novel kind of 'moral economy', which requires participants to adhere to specific technical and social standards on the production and 'open' exchange of biological parts and data (Frow and Calvert, 2013b; Frow, 2013). However, there remains considerable diversity in how community members negotiate the 'value' (Frow, 2013) of the parts-based approach to synthetic biology in relation to intellectual property and community building, and in their everyday work and career trajectories.

The moral economy of the parts-based approach, particularly as it refers to the sharing of parts in registries, is most clearly manifest in the context of the International Genetically Engineered Machine competition. The iGEM competition was set up by a small group of synthetic biologists at MIT in 2003, and has rapidly grown into a global contest, including over 4500 participants in 2014. It challenges teams of undergraduate students to construct novel biological systems by creating and using new and existing DNA parts that adhere to particular material and technical standards. The so-called, 'BioBrick™' standards are named so as to imply that clicking together standard biological parts will be as easy as is clicking together Lego® bricks. To be successful the teams must also conform to values of 'open access', by sharing their parts online, and of interdisciplinary collaboration in addition to a host of other norms embedded in the competition's reward structure (Balmer and Bulpin, 2013). In doing so, iGEM is putting in place distinct community boundaries for synthetic biology by disciplining the new generation of bioengineers in particular ways (Bulpin and Molyneux-Hodgson, 2013; Cockerton, 2011). iGEM has become a key 'community-building device' (Molyneux-Hodgson and Meyer, 2009: 139) and is responsible not only for recruiting participants and embedding specific values and research agendas within the community but also for establishing and populating key infrastructures, such as the 'Registry', as well as circulating tools and materials such as standardised biological parts.

Molyneux-Hodgson and Meyer (2009: 130) point out that iGEM operates as one of a set of 'devices' – alongside journals, conferences and 'success stories' – that are constructing synthetic biology as a 'universal collective enterprise'. The same kinds of norms, celebrations of success stories and promotions of community can be seen at the Synthetic Biology X.0 Conference (SBX.0). Since SB1.0, in 2004 at MIT, every few years academics from a range of disciplines and from across the world

have convened to discuss the status of the field and to share their findings, forge networks and discuss the future. In the context of governance,
one can also highlight community-building mechanisms, such as policy
initiatives, workshops and funded networks, all of which are operating
at local, institutional, national and international levels (Kearnes, 2013).
In this way, synthetic biology communities are taking shape across
different locales. Nonetheless, sociological studies have tended to focus
on the 'global' community-building mechanisms of a small core group
of synthetic biologists committed to the parts-based approach, and we
have relatively little empirical and theoretical work concerned with how
these larger networks of scientific, governance, industrial and material
actors filter into, and are fed by, more peripheral actors. A focus on the
periphery is central to our account.

The case study: synthetic biology and the water industry

Our ethnographic study in this book concerns how we tried to collaborate with a group of academic engineers and natural scientists, who
were concerned to enact synthetic biology at the university in which we
were all based. We did so as part of a funded research project designed
to explore if synthetic biology could be made relevant to the problems
currently facing the UK water industry. As was written in the grant
application:

> Deploying synthetic biology within the water industry has the poten-
> tial to generate more sustainable solutions (beyond those currently
> based on chemical or physical techniques) to address some of the
> key challenges facing the water industry, associated with continuing
> to maintain the provision of safe water supplies, hygienic sanitation
> and good environmental management in a sustainable manner in
> spite of increased urbanisation, aging infrastructure and changing
> climate conditions.

The research project was funded as part of a call from the Engineering
and Physical Sciences Research Council in the UK, the aim of which
was to examine the feasibility of cross-disciplinary activities. Bringing
synthetic biology into the water-engineering context was a good match.
The specific technical challenges posed by the project regarded two
imagined applications: 1) a bacterium engineered to act as a biosensor
for detecting the presence of pathogens or other contaminants in
water sources, such as rivers, reservoirs and so on, which could be used

by process engineers in the field to test water safety quickly and reliably and act on the information provided; 2) a bacterium engineered to create a biofilm on the inside of sewerage pipes that would make for a smooth surface, reducing the friction of sewage passing through pipes, and so lowering the energy required to pump sewage through the vast infrastructure. Although the engineers had developed these ideas and put them into the grant application, the nature of it being a 'feasibility study' meant that they could change them once the funding was awarded. The project timeline embedded 'industry days' that would allow for academics and water company actors to think more deeply about the kinds of applications that synthetic biology might provide. So there was some flexibility in how synthetic biology might find its test cases for innovation in this context.

Academic researchers in the team came from the social sciences, molecular biology and chemical, environmental and water engineering. The research project thus made for a triple cross-disciplinary feasibility study. First, synthetic biology is often conceptualised as being inherently interdisciplinary, in the way that its proponents tend to define the field by emphasising the use of engineering epistemology in the design and construction of biological entities. Second, the use of synthetic biology in the water industry manifestly brought actors from across different disciplines, including, for example, at least four different kinds of engineering, together with industry. Third, the design of the project was intended also to cross the disciplinary divide between the natural, physical and social sciences.

In the early grant-writing meetings, the engineers in the group were clear that the problems under consideration were not simply 'technical' but rather involved a number of 'social dimensions'. Most keenly felt by our colleagues in this regard were the 'barriers to innovation' that they saw in the water industry, based on their long-term relationships with water company Research and Development (R&D) Managers. The barriers were understood at this stage to include long-standing concerns such as: the water industry's 'conservative attitude' to innovation, the perceived 'over-regulation' of the industry and the 'public's ignorance' of water treatment services and of their cost. To this was added a new concern and barrier to potential innovation in the form of the possible 'public fear of synthetic biology'.

Our equal funding as sociologists in the project was in part justified on the basis of these imaginations of the 'social problems' and 'barriers' that might be encountered. Posing the questions we would explore in terms of social problems and barriers was something that we accepted as part of trying to make our work make sense to scientists and engineers and to win funding. Making use of the concerns, language and criteria of collaborators and their funders was, in our view, necessary to begin with but also subject to ongoing reflection and change. When writing the proposal we and our engineering colleagues were comfortable with the idea that orienting to barriers was an acceptable framing for the research council, but that it did not fully capture the position that we would be taking once we ventured into the field. Indeed, Susie had more long-standing relationships with some of the engineers, and had discussed topics such as scientific practices, governance and so forth as part of her previous work with them. This meant that a few of the people with whom we worked were primed for us to be challenging concepts like 'barriers' and were aware that we would be critically engaged with the constitution of synthetic biology, not just casual adopters of its norms and promises. As one of the engineers in the project put it early on in our interviews and observations:

> Andy Balmer: What do you see as the role of the social scientists in this project?
>
> Academic Chemical Engineer 1: That's difficult because, uh, I don't know. It depends what you mean by social scientists. I know the type of social science that Susan does, for example. I understand how that fits in. I don't know how broad you want me to answer the question? What do you mean by social scientists? Because if you're studying, if one is studying the formation of communities or the way that new communities work or whatever, that's obviously relevant in a young area like synthetic biology. ... That's an important role. As someone working in the field, I don't know how the field works exactly. Right? So that would be in a self-reflective way. [When we applied for the previous funding for a Network in Synthetic Biology] it did seem important to try to understand all aspects of that area. How people were working with each other, how could people best work with each other? What actually is it that we're doing as a community? I think those things are important. I think it's still new enough that I imagine that that's a thing that you guys are finding interesting. I'm finding it interesting too, because it's not something I'd ever thought about before.

As this quotation suggests, some of those we began to work with were sensitised to the kinds of issues that we might be exploring and under-stood that 'barriers to innovation' and so forth would be topics that we would examine and encourage reflection upon, rather than passively adopt as the problem to be treated. However, on paper, at least, we were still positioned in such a way that our job was to be about determining how feasible it was socially to use synthetic biology in the water industry as regards various 'social dimensions'. In practice, these concepts were still a part of a number of our colleagues' day-to-day framings of our work and of the challenges that they faced, particularly perhaps when they interacted with other natural scientists and engineers. This was the case even for those with whom Susie had previously worked. So whilst our colleagues sometimes adopted more nuanced understandings of the social nature of science, and reflected on scientific practices, they also nonetheless made use of, to us, more troubling notions regarding barriers and public ignorance and fears of genetic modification (GM). Thus there was a 'mess' of different concepts and positions jumbled up in the project, and often jumbled up within individual's accounts. For example, one water engineer described why he was interested in the research the team would be conducting, using a mix of different positions on the 'social side':

> The main thing that is of interest to me, that I get excited about, is the social side. Although I've probably got this wrong and probably Susie would hit me. What are the barriers and challenges to actually getting synthetic biology accepted? Actually, what would stop this ever being released into distributions systems, into the sewerage system? All this stuff. Where you've got modified organisms going out into the wild, what are the barriers, the challenges, policy issues, social perceptions? All that stuff. I guess there's probably a bit of what is the research process or field itself. But for me that's not a headline task. I understand it is for you and other people. (Academic Water Engineer 1)

It was clear early on that we would have to work within a mix of different and changing positions, both at the group level and with individuals in the project. Once the funding was awarded, we sought to open up the question of how a sociological analysis might contribute to, and expand on, the understanding of 'barriers' to the success of synthetic biology, and to rework and refine the concepts and practices that underwrote this vision.

Much of our work, from mundane encounters in project meetings to published papers after the grant had concluded, was about trying to alter how collaborators used ideas of 'social dimensions' and 'public fears' in their accounts and in how they constructed their work. Significant amounts of our effort at working alongside natural scientists and engineers in the project were spent on trying to change how the problems with which they were concerned were conceptualised. The success or not of this attempt at collaboration remains open to question, and is something that we return to throughout the book as we try to understand how our work figured in this project and the implications for the position of social science in synthetic biology more broadly.

What is more certain, however, is that we were a prominent feature of the academic collaboration from the design stages of the project right through to its end. As three sociologists working at different points of our careers – one a senior academic, one a newly minted postdoctoral research associate and one a Ph.D. student – we were able to engage with different actors at different levels of academic work. Collectively, we have been witness to the project's conceptualisation, implementation and conclusion. We have also tracked some of the project's legacy at the institution and beyond. So, our empirical data is collected from across a broad range of sites and over a number of years. The project was designed and funding awarded in 2009, but there were a number of important precursors, which date back to at least 2005 and which shaped our enrolment into the field. The legacy of the work we report on continues to develop and as we write this in 2015, we are all still involved in trying to collaborate with scientists, engineers and others in the context of synthetic biology. However, the bulk of the research was conducted between 2010 and 2012, through a series of planned interactions and specific social relations built into the structure of the grant.

The research project became the level of social organisation at which much of our ethnographic endeavours operated. We engaged with actors as they moved into contact with the project in various ways, whether through direct participation in experimental work or through attending a meeting, workshop or conference. The people who came into contact with our project worked at various levels of social hierarchy, whether these were academics, including undergraduates all the way up to professors and university managers; industrialists, including process engineers in a sewerage works up to corporate executives; governance professionals including volunteers in public regulatory bodies up to members of parliament; or members of various publics, including people in a street observing an experiment taking place up to members of campaign groups. They were all also primarily in the UK, but human actors did travel across international borders, whether through invocation in the local situation or literal contact during conferences, research trips and so forth.

The non-human actors in contact with the project were even more diverse, from Petri dishes to policy frameworks. These, too, moved into and out of contact with the project over a number of years, were organised at various levels of sociomaterial hierarchy and crossed international borders. They were anything from the microscopic to the climatic, from the mundane to the extraordinary. Moreover, they all, in various ways, exerted their own forms of agency in the sociotechnical production of the project.

Of course, we didn't study every actor in great detail, nor were the research questions that each of us held particularly overlapping, but instead we pursued lines of investigation by following actors more or less closely as they became more or less interesting as we sought to understand how synthetic biology was being pieced together in the local situation. Some loci of concentrated study included the project's experimental and bureaucratic practices, the iGEM competition and industrial sites of clean and waste water management.

Our study of a project focused on water industry problems represents a novel and powerful approach to the study of synthetic biology in three ways. First, synthetic biology has largely been studied in isolation from the contexts of sociotechnical research in which it is being positioned. We take a deeply situated approach, investigating the emergence of synthetic biology in our academic water research and industrial engineering field sites and in our colleague's careers. We locate the emergence of synthetic biology within pre-existing sets of sociotechnical water problems, academic and industrial practices, water governance regimes and related political, economic and power relations. This allows us to examine how synthetic biology is emerging from within the world and not miraculously from out of a vacuum or from a teleological progression of scientific enlightenment. We take an ambivalent approach to the definition of 'synthetic biology', seeking to understand it from within this given situation rather than to promote any particular definition. As such, we are able to better understand how synthetic biology takes the shape that it does by reference to the practices in which it is enacted. In this regard, we are studying synthetic biology 'in situ'.

Second, our approach focuses as keenly as possible on the everyday lives of academics and industrialists engaged in making synthetic biology. So much that has been written about synthetic biology emphasises what are seen to be unusual, novel or superlative features of the field, whether on the promises of curing malaria worldwide or creating a sustainable biofuel that will save us from climate change, the fear of terrorists making biological weapons, the utopia in which biological design becomes mundane and everyone can create bespoke microorganisms in their kitchen or questions about what constitutes life and whether scientists are 'playing God'. Instead, we emphasise the routine features of academic work in making synthetic biology a reality, from writing grant applications, trying to get bacteria to behave in a laboratory, working on long-standing banal problems like leaky pipes or sewer sedimentation or just trying to get a promotion or stake a claim to an intellectual territory. Rather than elevating synthetic biology to the status it regularly receives as a promissory technoscience, one of the 'Eight Great Technologies' (Willetts, 2013) in line to save the UK economy, we try instead to make it as boring as it actually is in the daily reality of doing such work.

Finally, the limited numbers of ethnographic studies of the field so far have been conducted at the *core* of synthetic biology. Significant work has been done at the major US Synthetic Biology Engineering Research Centre (Rabinow and Bennett, 2012) and within the various projects based at Imperial College London, for example in the Centre for Synthetic Biology and Innovation (CSynBi) and the Synthetic Biology Innovation and Commercialisation Industrial Translation Engine (SynbiCITE) (Cockerton, 2011; Finlay, 2013; Marris, 2015). Our project took place outside this core set (Collins, 1988; Collins 1992 [1985]) of people, practices and universities. Instead, we examined how work was being conducted at the periphery of synthetic biology's emergence and concretisation. One of our key questions thus concerned how the emerging values, practices, imaginaries and so forth at the core of synthetic biology were made sense of, adopted, transformed or resisted at the periphery. Invoking a distinction between the core and the periphery of synthetic biology warrants further elaboration, particularly since it might appear to pose a challenge to our claims regarding a disavowal of any specific definition of the field.

The core and the periphery of synthetic biology

Having argued that we will not be adopting a particular definition of synthetic biology and that we are instead interested in how we and our colleagues worked to enact 'synthetic biology' within the local situation of our collaboration, it may seem antithetical now to make a distinction between the 'core' and 'periphery' of synthetic biology. To some degree, it is unavoidable that we do draw a ring around what we will count as being relevant to our study of an emerging technoscience. We cannot, and have not, studied everything that has been brought under the description of synthetic biology. We have had to make decisions about what kinds of things we will study. This has steadfastly not involved producing our own description of what will count as synthetic biology and what will not. Rather, we have sought to bring into our view those things that are being claimed as synthetic biology from within the local situation of our engagements with our colleagues. From engaging in this work it became clear that there were various phenomena that might be considered to be operating from the 'core' of synthetic biology and various other phenomena that were not, but which nonetheless involved claims to the term.

We argue for this distinction based on a number of observations. First, the idea of core synthetic biology practices and concepts is a powerful category performed by many of our collaborators themselves. Some of them distinguished between, and generally strategically so, various practices and concepts as being core to the field. For example, one colleague talked to us about how the concepts of standardisation and testing in engineering were core to synthetic biology practices in order then to be able to claim that certain of the group's long-standing practices of working with bacteria counted as synthetic biology. At other times, the term was used to refer to a core set of people, generally having to do with the development of synthetic biology at what are understood to be key institutions in the UK and the USA.

This brings us to the second dimension of the distinction. There are, even with a cursory glance, institutions that are more obviously engaged with synthetic biology through their explicit use of the term in their researchers' job titles, awarded grants and public claims or in the names of departments, institutes or research groups. Such activities are spread out across various institutions in the USA, the UK and Europe, but they cluster in particular around a subset of universities, including the University of California, Berkeley; Massachusetts Institute of Technology (MIT); Imperial College London; Edinburgh University and the University of Cambridge. More recently the group of 'core' institutions has begun to expand, as new funding programmes have awarded large grants to several universities (for example in the UK synthetic biology research centres) to finance the purchase of expensive automation, sequencing and testing technologies, and to set up synthetic biology research groups. In this regard, actors engaged in practices that they term synthetic biology might compare themselves and their work, more or less favourably, to those researchers and the research being conducted at the core institutions, in order to judge their progress towards the goals of synthetic biology. We saw this happening at the level of established academics in our project, but also within the undergraduates in the iGEM team, who emphasised how teams from Cambridge, Imperial and MIT, for example, were at an advantage in terms of resources and synthetic biology expertise.

These clusters of activity thus also map geographically onto long-standing inequalities in the distribution of research funding more generally. Large universities with strong reputations in science and engineering (for example, Cambridge, Imperial, MIT, the University of California) are more able to enforce particular norms around the development of the new field by branding their projects in particular ways and by lending their reputation to the kinds of standards and norms with which they work.

Again, this connects to another distinction between core and periphery, namely the authority publicly to define what synthetic biology is. It is these core institutions that tend to be the homes of actors who such performative statements. For example, in one of our 'artefacts from the field' above, Professors Richard Kitney and Paul Freemont, both working at an RCUK-funded synthetic biology research centre, at one of the core institutions, Imperial College London, were explicit in their demarcation of the field from other practices. Indeed, they also drew a line around those activities that may be counted as doing synthetic biology and those people who may be counted as being part of the community. Such a claim is warranted, in part, by their positions not only within CSynBI but also at Imperial, and by virtue of their prominence in publishing in this area and by having produced the first UK textbook on the field. Moreover, both professors are public scientists, in that they are both regularly interviewed in the UK media when synthetic biology is discussed. Both have also held powerful positions in governance in the UK and internationally, which leads us to our penultimate point on the distinction.

Core synthetic biology is also organised through the connections that actors at core institutions have within governance. They can shape the field in particular ways, supporting distinct ways of working, for example, by expressing preferences over public and private approaches to DNA repositories. As a further example, Prof. Kitney was one of the main actors driving forward the funding regime for synthetic biology in its early years and has played a central role in recruiting members of the UK parliament, international industries, and regulatory and funding agencies into the network of actors working to constitute a particular vision of synthetic biology in the UK and abroad.

Finally, actors both at the core and the periphery of synthetic biology help to demarcate what is to count as synthetic biology by drawing attention to exemplars of the field. Such exemplars are often tied to particular core actors at core institutions. Perhaps the most significant case is that of the synthetic fabrication of a precursor to artemisinin, an anti-malarial drug, developed by researchers at Synberc. The research conducted on artemisinin fabrication is regularly held up at conferences, in the media and in everyday talk as examples of the success of synthetic biology, of its values and practices, and of the promise of the field for solving global sociotechnical problems.

In summary, there are six main factors that we think justify the demarcation of core from peripheral synthetic biology in our account:

1) Actors from our project singling out core practices, concepts and people;
2) the connection between key actors and particular institutions;
3) the accretion of resources at those institutions and around those key actors;
4) the ability of those actors to then make public declarations about the definition of synthetic biology;
5) the connections between these actors and those in governance; and
6) the citation of exemplars and success stories from the field.

That being said, by choosing two terms for our demarcation we perhaps become over-reliant on a binary classification, which is not reflected in the messiness of synthetic biology and sociotechnical co-productions. We acknowledge that there is much entanglement, and that core or peripheral practices, concepts or people should not be objectified as such for all time. We use the heuristic for our purposes in this book, but do not expect that the demarcation will be relevant to every study of the field. Indeed, our emphasis in this book is on the mess of everyday life. But, as we have already pointed out, it is from within this messiness that such distinctions have been made. So we use the terms of core and periphery not only as an analytical category but also as part of our data in order to understand how classifications like key actors or core practices are used to construct the field of synthetic biology in particular ways. We adopt the distinction to the extent that it proves useful but we hold it up for future scrutiny, nuancing or abandonment as becomes useful or necessary.

Alongside this conceptual apparatus, three further themes have provided ways into the world of synthetic biology and have given a means to shape our contributions to science and technology studies more generally.

Sensitising concepts

Our study of synthetic biology is framed within key STS literatures that are intended to sensitise readers to a particular way of viewing the conduct of scientific work and that will allow us to emphasise some crucial research questions that have yet to be posed.

Changing practices in everyday life

As indicated by the subtitle of the book, our work is about practices and specifically about change. It emerges from ethnographic research, including interviews, observation, participation, conference attendance, focus groups, documentary analysis, creative methods and self-reflection, conducted during a funded research project. Our everyday practices as STS researchers were focused on the everyday practices of scientists and engineers across several locales.

The practice lens has an illustrious history within STS and conducting a 'lab ethnography' study remains for many a rite of passage into STS research life. Invoking Latour, to 'follow the actors' (Latour, 1987), we could not confine ourselves to laboratory practices in the sense of predetermined sites of scientific work, but rather we needed to travel wherever the actors took us. Again quoting Latour (1996: 46):

> We sociologists have to drag ourselves around everywhere...Our terrains aren't territories. They have weird borders.

Empirically grounded approaches to scientific practice characterised the development of the field of science studies and so informed the kinds of claims made about the nature of science and technology. Complexity, partiality and diversity of scientific practices have become major themes for science studies scholars. Ethnographic investigations into scientific practice often study research within a particular location (for example, Latour and Woolgar's (1986) endocrinology lab); or attend to a specific experimental topic (for example, Collins (2010) gravitational wave studies). Practices, across multiple sites, are manifest in multiple forms, whether in their embodied performance, verbal or written citation, their conceptualisation or problematisation. Amalgams of practices construct what is and is not acceptable for a 'community of practice' and with implications for how groups of actors make sense of the world and act in it. Traweek (1988: 9) describes shared ground, 'the daily production and reproduction of what is to be shared,' as the basis of practice and not 'some a priori norms.' In this regard, studying scientific practices highlights the constant work required to maintain communities and sameness, whereas the term discipline usually denotes a fixed picture of the conduct of science (although see Bulpin and Molyneux-Hodgson, 2013 for a counter view). Alongside practices, terms such as epistemic cultures and communities have been created to point to how practices and boundaries are more 'open to debate' and 'dynamic' (Meyer and Molyneux-Hodgson, 2010) than disciplines as traditionally conceived.

Practices make for an appropriate form of approach to the construction of synthetic biology and its contingency. It is changes in practices of design, experiment, construction, manufacture and industrialisation that synthetic biology proponents tend to highlight as being constitutive of a new field. Attending to everyday practices, in particular, helps to take account of how practices are always undergoing change and adaptation according to the specificity of the situations in which they are performed, and of the shifting rhythms of everyday life (Blue, 2012). The practices of synthetic biology have yet to be stabilised. They remain open to debate and contestation, and, certainly, the promised reconfiguration of science and industry is far from realised. Synthetic biology, as we will show throughout the book, is emerging and consolidating from within existing practices that have their own specificity and rhythms. On this note, we tackle the complex entanglement of emerging and established practices of governance, innovation and knowledge production by emphasising the everyday work of academics, industrialists, students and professionals, all engaged to various degrees in the realisation of synthetic biology. It is a rather messy picture.

In orienting to synthetic biology through the lens of practices we can be attuned to the mess and complexity of everyday life in academic, governance and industrial spaces as actors work to bring about change. We explore how an idea of the future, in which the use of engineering epistemology has helped to bring about changes in how we industrialise microorganisms, is connected to the everyday work of some of those actors experimenting with synthetic biology practices in the present. This means we have to attend to how the past, present and future are entangled in this emerging field.

Promises and imaginaries

As we have outlined, synthetic biology is *the* new technoscience. It has promised to tackle global sociotechnical challenges in the areas of food supply and waste, fuel and energy crises, health and illness and climate change and pollution. The ways in which emerging fields, and the institutions that support them, are orientated towards the future has become central to STS accounts of changing practices. A key idea is understanding how 'expectations' are generated and how they are used to mobilise resources (Borup *et al.*, 2006) in the present.

This sociology of expectations has been explicitly used to critique biotechnologies, for example regarding the ways in which hope and hype are used to bring about investment in particular bio-innovations (Brown, 2003). Specific technologies that have received a diagnosis of being over-hyped include pharmacogenetics (Hedgecoe and Martin, 2003) and biobanks and stem cells (Brown and Kraft, 2006; Eriksson, 2012; Martin *et al.*, 2008), amongst others. There is already an emerging literature on the ways in which expectations in synthetic biology are being constructed (Hilgartner, 2015; Marris, 2015). The sociology of expectations acts as a frame that allows us to view how the future is performed in order to bring about synthetic biology in particular ways.

For example, something as simple as the use of the term 'synthetic biology' in academic, industrial and governance situations is part of how actors in these situations assemble problems in relation to different imagined futures. Making promises about the future of synthetic biology helps to assemble networks of actors now, in order to try to bring about such a future in which modified microorganisms and novel biological entities are commonplace. This affinity for future-making – through use of speculation, anticipation and hype – helps to align the scientific enterprise with prevalent neoliberal ideals (Lave *et al.*, 2010; Ylönen and Pellizzoni, 2012). Synthetic biology epitomises the kinds of co-productive relationships (Jasanoff, 2004) understood to be at the heart of the governance and production of knowledge. The science of synthetic biology and the imagined society in which it plays a role are being made together. As such, the past, present and future of synthetic biology are entangled with social order more generally.

In the vein of co-production, synthetic biology makes for an interesting case in how epistemology is being assembled in relation to multiple more or less complementary or conflicting norms and ways of ordering the community and society more broadly. As we described above, there are conflicting accounts of how synthetic biology's approach to engineering should be tied to norms of sharing and community, which relate to the mechanisms through which different actors foresee the industrialisation of microorganisms and what kinds of socioeconomic systems might therefore be brought about. Taking just the 'parts-based' approach, it is clear that several different claims regarding the future are being used to assemble actors into particular networks in the present. Some are organised in relation to the sharing ethos, which is particularly visible in the case of iGEM, whereas others are organised more explicitly in relation to economic goals and are adopting more traditional patenting approaches, for example around the six synthetic biology research centres in the UK. These visions of the future also have implications for how the research councils and government departments operate in the present. Some visions of the future of synthetic biology are very much entangled with visions of the economy and of nations more broadly. Indeed, synthetic biology has become one territory in which states are understood to be in global economic competition. The imagined future of the country that can industrialise biology fastest and with most ingenuity has become a vision through which contemporary state actions are organised. Such entanglements between imagined futures thus represent 'sociotechnical imaginaries' (Jasanoff and Kim, 2013) in which the state and science are co-produced in the present by virtue of their imagined entangled futures.

How sociotechnical futures are imagined in order to bring about present networks of actors at the nexus of academia, industry and governance is thus vital to understanding the ramifications of synthetic biology in given situations. However, this is not only about human actors. The use of promissory narratives and sociotechnical imaginaries both depends and redounds upon the constitution of materiality. What form microorganisms take, through the use of particular practices, organised in relation to particular imagined futures, is an issue of central importance not only to synthetic biologists but also to STS scholars interested in how objects come to be.

Enacting ontologies

STS has long taken an interest in objects and things. Several of its key theoretical developments have been explicitly concerned to take things seriously. Whether it is bicycles, bridges or bullets, things have been given their say in how practices develop, change and persevere. Such a long-standing interest means that we now have many accounts, from a diverse range of contexts, of how it might be that materiality figures in the co-production of science and society. Concerning 'things', much of the STS engagement with synthetic biology has emphasised the emergence of novel practices for working with biological materials and how the primary goal in the field is to use engineering epistemology to reduce the agency of the biological realm and so to make it more predictable and easier to industrialise (Calvert, 2012; Campos, 2012; Finlay, 2013; Frow, 2013; Mackenzie, 2010; Molyneux-Hodgson and Balmer, 2014; Schyfter, 2013b).

As synthetic biologists have sought to take ever greater control of what they term 'genetic circuitry', a common refrain can be heard from the unconverted. Those more critical molecular biologists often point out that bacteria resist our attempts to manipulate them. Such contestation over agency is music to the ears of an ethnographer. However, it also shows that synthetic biologists themselves must explicitly orient to the wilfulness of things as they go about creating new practices for their control, and that this is part of how they are constructing the field alongside dissenting voices. How this pans out in everyday practice is one area we illuminate in the book.

On this issue, STS researchers have documented the entanglement of objects and different ways of knowing the world, which is to say that they have shown how epistemologies are extended, fitted together or held apart, or erupt into contestation through relation to the status of their objects. Latour's (1987) ironically well-travelled concept of the 'immutable mobile', for example, examines the ways in which objects have action at a distance and so help to carry facts across different situations. He shows that this depends upon the tight networking of various actors and so demonstrates how objects, networks and ways of knowing are mutually constructed. The equally well-known concept of 'boundary objects' is another case in point, since they are precisely those kinds of things that allow for 'different groups to work together without consensus' (Star, 2010: 602). In both of these concepts, there is a sense of the management of sameness, but also of difference and change.

STS scholars have often highlighted how consistency is something that is accomplished not given. Research has powerfully evidenced contestation and variation even in those things that we generally take to be fundamental and fixed. So, rather than seeking to produce its own classificatory schemes, STS has instead concerned itself with how such schemes are made, altered and discarded, and with what relation to broader projects of ordering society (Bowker and Star, 2000; Waterton, 2002).

This can be seen in the context of synthetic biology, where Calvert (2010), for example, has shown that what constitutes the 'natural' is a receding horizon and is articulated through its opposition to other terms, like 'synthetic' and 'artificial'. This articulation of dynamic oppositions is a part of how synthetic biology is being made within science and governance. How synthetic biologists seek to distinguish the field from previous forms of biotechnology is also an issue of classification, as is the repurposing of engineering terms like 'part' and 'chassis' for biological contexts.

Examinations of classification and blurred boundaries have meant that STS has become more or less comfortable with slipperiness (Law and Lien, 2012), mess (Law and Singleton, 2005) and multiplicity (Law, 2002; Mol, 2002) in one form or another. Developing out of this trajectory, the recent 'turn to ontology' (Woolgar and Lezaun, 2013) in the social sciences has sparked debate in the STS community regarding the ontological implications of our messy descriptions of objects and practices.

The argument regarding ontology in the STS literature seems to be that when we produce messy and multiple descriptions of the co-production of practices and things, these are not merely a reflection of different perspectives on the same thing. Instead, things, like microorganisms or biological parts, are now to be understood as multiple. We should see them as being *enacted* differently through different practices. Summarising this distinction of enactment – more familiar concepts having generally to do with perspectives, attitudes, beliefs and the like – Woolgar and Lezaun (2015: 463) contest that there is:

> an important difference between the notion of 'social construction' and the references to 'enactment' that can be found in much of the STS literature on ontology: the former describes social processes that result in durable realities, while the latter describes practices in the here and now that produce ephemeral effects – effects essentially coextensive with the practices that create them.

Multiplicity, then, is not just to apply to viewpoints, attitudes or ways of knowing but also to ways of doing the world (Mol, 2002; Law and Singleton, 2005). Further, the emphasis here on doing is precisely the point – that things get done, they are performed, differently. Throughout the book, we explore some of this ontological and epistemological mess in the shifting terrain of synthetic biology as our colleagues worked to make the field in their local environment, and how this related to the ways in which things like 'barriers', 'bacteria' and 'bodies' were enacted differently across shifting situations and over time.

The organisation of the book

As we have described so far, the book takes the empirical study of a synthetic biology project at the nexus of academia and industry as its focus. It is a deeply participative sociological investigation into how a very contemporary mode of technoscience that promises 'revolution' in current social and technical orders is emerging as a set of situated practices across different locales. In particular, we want to interrogate what kinds of re-orderings, realignments and adaptations have been necessary to make synthetic biology 'fit' into the everyday rhythms, routines and spaces of academic and industrial labour in our given situation. Although we can identify such points of 'fit' and 'comfort', we are also attuned to places where things did not fit so well, the more uncomfortable couplings and the resultant misalignments that occurred.

We use the project to think through our own methodological and theoretical engagements with synthetic biology as well as to explore the transformative objectives of a new technoscience in which identities, practices, relationships, modes of training and institutional forms are all part of the experimental mix. We are particularly interested in how the hopeful promises of synthetic biology are becoming rooted in the sometimes hazardous practices of the everyday. We explore how the mundane and ordinary spaces, practices and problems of academic and industrial work respond to the imaginations and rhetoric of synthetic biology's discourse and emerging practices. In examining how synthetic biology is developing 'in situ', we probe the dynamic and multiple material, social and temporal relations that are constituting synthetic biology as an everyday practice and how these ramify across many levels of entanglement.

In summary, we aim to contribute to the literature on synthetic biology by understanding how its terms and practices spread and are made sense of in different situations. We extend existing analyses of synthetic biology by taking a mundane, everyday approach. In doing so, we expand upon the emerging research regarding the multiplicity of objects and how ontologies are enacted through situated practices. Our examination of how synthetic biology was enacted in our local situation was importantly framed by the context of water engineering, and of our colleagues' attempts to reconfigure academic and industrial practices to fit together. As such, we are able to comment on the shifting relations between knowledge production and industrialisation in contemporary sociotechnical politics, particularly as regards to how boundaries are drawn and barriers enacted. Finally, by virtue of our attempts at collaboration, we also contribute novel findings regarding the emerging 'turn to collaboration', the development of 'post-ELSI' methods and the move to get beyond 'ethical, legal and social implications' (see Coda 1) for working across the natural, physical and social sciences (Balmer and Bulpin, 2013; Balmer *et al.*, 2015; Rabinow and Bennett, 2007; Rabinow and Bennett, 2012).

Rather than pursuing a particular timeline in our description of the project, we organise the book according to three 'slices' through the data, taking 'barriers,' 'bacteria' and 'bodies' as a way to explore the broader STS themes we outlined above. These three slices are interspersed with codas. Rather than a conventional outline of methods and description of characters and field sites, we use the coda after each main chapter to document some of how we engaged with our colleagues in the project. In this regard, these interventions in the narrative of the analytical text are redolent of the movements we have to make as fieldworkers, moving in and out of the field on a regular basis, spending time in collaborative activity with people whose practices are far removed from our own and then coming back to home ground. Thus, we use the codas to reflect on the roles that we, and social science more generally, played in the project.

Immediately following this introductory chapter, then, is the first coda, which describes the back story to our, and other sociologists', encounters with synthetic biology. The field of post-ELSI practices is laid out and our work situated in relation to this. We emphasise the importance of these codas for our reflections on our collaborations with natural scientists and engineers in relation to the themes we explore in the main empirical chapters.

Chapter 2, 'Barriers', recounts aspects of the imagination and enactment of barriers to innovation in the water industry, situating this within longer-term projects to reconfigure the relations between academia and industry. We examine how our colleagues worked to make synthetic biology a test case in the potential for novel, high-tech solutions to global problems in the context of water services. Ultimately, we explore how 'barriers' to change are conceptualised and enacted in everyday life, and how change in practices in one time or place is entangled with others at multiple levels of social order.

Coda 2 tells of 'Brokering' – the negotiation of the in-between-ness of situated practices, in order to explore how we sociologists were situated at the nexus of industry, academia and publics. In particular we explore how our expertise in 'social issues' was negotiated and with what consequences for our collaborative endeavours.

In Chapter 3, 'Bacteria', we explore the ways in which practices of working with bacteria were variously constructed, contested, shifted and consolidated as actors in different sites worked to adopt or adapt existing practices and to overcome barriers to their desired ways of working. Central to this concern is how bio-objects like bacteria are enacted differently depending on variations in practices over time and across spaces. We situate this work within the everyday lives of our colleagues, regarding how they worked on problems, tried to develop their careers, changed how their department was situated in the university and managed their relationship to core synthetic biology activities.

Forms of sociological disruption are then discussed in Coda 3 on being 'Critics on the inside'. We examine how working critically with the emergence of synthetic biology from within a collaboration caused problems for the consolidation of the field in our local situation, meaning that our methods became a potential problem for the success of the project as our colleagues envisaged it. Ultimately, our critical engagement with the field had to be internalised, since the disruption became difficult to maintain if we were to keep our collaborative relationships going.

'Bodies' are the subject of Chapter 4, the final empirical chapter. The chapter addresses the centrality of bodies and embodied relations to shifting ontologies and practices. We explore the idea of 'disembodying' the creation of novel microorganisms, as is crucial to the promises of automation in the field. We describe how attempts are being made to reconfigure bacterial, human and public bodies into new kinds of arrangement with each other and with the material, temporal and social organisation of different spaces. We also point to how existing enactments of such things might resist attempts to change them, and with what implications for the hopes of certain imagined futures.

Coda 4 centres on care, affect and nurturing as aspects of the 'Reciprocal Reflexivity' that we developed as one strategy for engaging collaboratively with our colleagues in the natural sciences and engineering. We consider how social science figured in the iGEM competition and what it was like for the iGEM team to do 'human practices' research whilst trying to engineer a biosensor for cholera.

In Chapter 5, we pull out and further develop some key findings of the book by slicing our investigations into synthetic biology according to ontology, failure and time. We find that these concepts are useful in understanding synthetic biology, but also in how practices change more generally. As regards ontology, we explore how we and our colleagues struggled to get new ontologies to stick or to unstick existing ones, examining some of how difference and sameness were choreographed in the project. This leads us into failure, where we explore the onto-affective dimensions of synthetic biology, examining the ways in which affective relations are managed both in the emergence of new ontologies and as actors in the field try to rework their practices around them. The time taken to overcome barriers, change ontologies, reconfigure practices, develop one's career and so on becomes the focus of our third theme. We examine how timing and tempo figured in our project and argue that time is crucial to how synthetic biology is being enacted, noting that time also has to be enacted through situated practices. We argue that considerations of ontology, failure and time help us to understand better the emergence of promissory technosciences.

We close with Coda 5 on how these three slices of ontology, failure and time also intersected in our attempts to collaborate. And, for a book centred on everyday practices, it seems right to conclude by reflecting on how things have changed for our own professional lives and academic practices as synthetic biology has changed over the years and our engagements with it have developed. So, we finish by reporting on where we are now.

Coda 1
Towards Collaborative Practices

Andy Balmer: Do these collaborative relationships factor into how you think about the success of the project?

Academic Water Engineer 2: Definitely. Once you've made collaborative links, if you make strong links, you'll keep them for years and years and years, and other things will spring out of it and you don't know what will happen.

AB: So building a relationship is a success in itself?

AWE 2: Well yes, and it is a very, very good way to build a relationship, working together collaboratively. It tends to be the stronger types of relationships that you build are the ones where you've got something to work on, rather than just endless discussions.

Since the first grants were awarded to projects explicitly concerned with 'synthetic biology', social scientists of various kinds have been involved with the making of the field. The UK research councils (RCUK) were early movers in the game of constituting synthetic biology as an interdisciplinary endeavour that would involve the social sciences. In 2007, the Biotechnology and Biological Sciences Research Council (BBSRC) and the Engineering and Physical Sciences Research Council (EPSRC) made funding available for the establishment of networks in synthetic biology, which also involved the Arts and Humanities Research Council (AHRC) and the Economic and Social Research Council (ESRC) in their development. The funding call was thereby framed to include ethics and stressed the need to handle 'social and natural issues' concurrently (Molyneux-Hodgson and Meyer, 2009).

This was, in large part, because governance actors, as well as some scientists and engineers, were keen to ensure that the economic potential of synthetic biology was realised. Their concern was that novel practices for engineering microorganisms would lead to contestation and controversy, which they understood to have been the primary causes behind the so-called 'failure of genetic modification' in Europe and elsewhere. In addition, the extraordinary promises made about the ability to quickly, cheaply and predictably manipulate life simultaneously opened up a raft of less positive 'implications'. For governance actors these generally had to do with fears about both the uncontrolled release of microorganisms into the environment and the possibility that synthetic biology tools would be used to create bioweapons.

Previous controversies around emerging technologies and promises of novel sociotechnical futures crystallised in a demand for more deliberate examination of the ethical, legal and social implications (ELSI) of science, particularly in the contexts of the Human Genome Project and later, nanotechnology. Social scientists and humanities scholars became tasked with addressing the 'ELSI' dimensions of new technologies and facilitating public engagement. They also played a significant role in creating novel governance frameworks that were mostly designed to remediate potentially negative consequences of technological innovation (Kaiser *et al.*, 2010). These frameworks were important in the first movements by the research councils to integrate social science into the funding regimes of synthetic biology.

This has meant that the majority of social science and humanities scholarship on synthetic biology conforms to an 'ELSI' mode of engagement, in that it is largely concerned with addressing potentially negative implications of proposed tools and technologies. These debates crystallise around issues of biosecurity (Kelle, 2009), biosafety (Schmidt, 2008), intellectual property (Henkel and Maurer, 2009) and an ongoing ontological concern with the definition of 'life' (Deplazes-Zemp, 2012). Such topics are also well rehearsed in the amassing grey literature on synthetic biology from national research institutes (Parens *et al.*, 2009), international research programmes (Schmidt *et al.*, 2009) and policy advisory groups (Balmer and Martin, 2008). The debate in public social media demonstrates an equally broad range of issues (Shapira and Gök, 2015).

Social scientists of various kinds have, however, expressed dissatisfaction with how their work is positioned within ELSI research programmes that shape these kinds of debates. In particular, they have criticised how ELSI practices encourage social scientists and humanities scholars to engage in 'speculative ethics' (Nordmann and Rip, 2009) about how technologies being developed now will impact on the economic, environmental and social future. Relationships established between natural and social scientists in those contexts have meant that ethical considerations remain separate from ongoing scientific and technical work (Swierstra and Rip, 2007). The science has proceeded irrespective of what the ethical and social science work has suggested might happen or should be changed. Indeed, 'ethical and social implications' tend to be understood by scientists as irrational, and, as such, as something to be overcome through exercises in good public relations and science communication.

Social scientists have thus been trying to renegotiate their positions in sociotechnical research projects and have sought to orient interventions more firmly towards the practices of technical innovation as it occurs. These more integrative attitudes have manifested in calls for greater upstream public engagement (Wilsdon and Willis, 2004) and the development of specific mechanisms for the governance and assessment of emerging technologies. These latter approaches include constructive technology assessment (Rip *et al.*, 1995), anticipatory governance and real-time technology assessment (Guston and Sarewitz, 2002) and 'responsible research and innovation' (Owen *et al.*, 2013). Concerns with 'opening up' the social and technical fabric of scientific practices and nurturing more reflexive forms of decision-making (Stirling, 2005) are common themes in this kind of work.

One such approach has been termed 'post-ELSI' (Rabinow and Bennett, 2009), in that it firmly distances itself from previous ELSI configurations and moves more explicitly towards a concern to develop 'collaboration' as an interdisciplinary method. Moreover, the 'post-ELSI approach' has emerged directly in relation to synthetic biology, developing out of Synberc through Rabinow and Bennett's (2012) experiments with novel forms of collaboration. 'Experiments' with the forms that interdisciplinary relationships can take have led some to propose that we push for a 'collaborative turn' in the social sciences (Fitzgerald *et al.*, 2014a) though this is perhaps most explicitly tied to research in STS (Balmer *et al.*, 2015; Fitzgerald and Callard, 2015) and anthropology (Rappaport, 2008).

These and other more long-standing efforts at interdisciplinary collaboration have contributed to the flow of social scientists 'upstream' into embedded or integrated positions from the very beginning of emerging technological developments. This has borne fruit in synthetic biology. Social scientists are now a mandatory part of political and scientific networks, for example in the UK Networks in Synthetic Biology, the

UK Synthetic Biology Roadmap Coordination Group and the six UK synthetic biology research centres. These upstream positions have helped to create an 'integrative field' in which experiments with sociotechnical collaborations can take place (Fisher *et al.*, 2015). Some of the sociologists embedded within such networks and research centres have begun to explore creative mechanisms for building greater reflexive capacity within synthetic biology practices (Balmer and Bulpin, 2013; Frow and Calvert, 2013b). Asking broader questions about the emerging epistemologies of the field and how they are entangled with wider political and economic imperatives is crucial here and characterises significant amounts of the STS research on synthetic biology that we explored in Chapter 1.

However, despite the institutionalisation and apparent integration of social science there is still a strong expectation coming from natural scientists and engineers that social scientists should be conducting research that looks more like traditional 'ELSI' frameworks. In other words, there is resistance from synthetic biologists to the idea that they should be concerned with reflecting on, and changing, their practices. Rather, they tend to envisage social science as being responsible for ameliorating the future negative implications of the technologies that they are developing. This can cause significant tensions between natural scientists, engineers and social scientists regarding how to proceed with collaborations, how to conceive of current work and possible futures, how to act responsibly and in what ways to engage with public actors of various kinds (Balmer *et al.*, 2015; Calvert, 2013; Calvert and Martin, 2009). Rabinow and Bennett's (2012) account of their experiences of working in Synberc, for example, paints a rather gloomy picture of how differing expectations can cause debilitating contention and antagonism.

Further examples of the difficulties experienced in working towards post-ELSI relationships have documented how power shapes the roles that social scientists can take with implications for the affective and emotional dimensions of everyday work (Balmer *et al.*, 2015). On this note, Fitzgerald *et al.* (2014b: 701) suggest that whilst common descriptions of interdisciplinary or transdisciplinary work tend to highlight notions of conciliation and integration, 'such projects may also be organized around some more subterranean logics of ambivalence, reserve and critique.' The body of research has begun to accumulate into a rather pessimistic diagnosis for the future of collaborative relations. However, there have been some more positive findings. In the area of synthetic biology, Balmer and Bulpin (2013) show that creative methods can be used to increase the reflexive capacity of science and engineering students, and Ginsberg *et al.* (2014) show that social scientists, designers, artists and synthetic biologists were able to invent novel approaches to the field when freed of particular institutional goals and empirical constraints.

But there remain questions about how these kinds of creative endeavours in experimental collaboration might contribute to the goal of producing more reflexive science, particularly when professional norms and power relations are more acute. As Balmer *et al.* (2015) point out:

> various positions and actions become differentially possible across space, types of engagement and over time. It is far easier to experiment with coproduction and induce reflexivity in the lab with a group of talented undergraduate students in an iGEM team than it is with a group of established professors of engineering and science during a meeting with cabinet MPs, civil servants and corporate executives.

In the project on which this book is based, we found that we encountered a number of these issues in trying to work towards post-ELSI collaborative relationships with engineers and natural scientists. Our access to the sites in which we worked was in fact shaped by the broader movement that we have described above regarding the institutionalisation of social science into synthetic biology. These changes helped to set up particular expectations and constraints. However, we each experienced them differently. This had, in large part, to do with the different kinds of work each of us was conducting as part of the project but also with our professional status and personal dispositions, and those of the people with whom we worked.

Susie, for example, already had good working relationships with several of the engineers involved in the project. This was because she had been working with them for some time on various other water engineering research projects. In addition, she was part of the synthetic biology research network, funded by RCUK, at the university, which brought together some of the actors that would play a role in writing the synthetic biology and water engineering grant proposal. In this regard, Susie was already a colleague to some of those with whom Andy and Kate later engaged.

Since 2008 she had a desk as a base for ethnographic work in one of the engineering department offices. Having worked with senior and more junior engineers for a while, she was quite friendly with some of them, so there were everyday personal and professional relationships that pre-existed the project. As a senior academic, Susie was then able to leverage the relationships and capital that she had with colleagues in the department when the grant was being put together. This meant that a fully funded postdoctoral position was written in for a sociologist alongside the natural scientists and engineers who would subsequently be employed. In this way, Susie played a significant role in the design of the water and synthetic biology grant proposal, ensuring that the social science dimensions were given proper weight. Her contributions therefore also shaped the broader interdisciplinary approach taken in the project. Her connections and status in the grant acted as a route in to the project for Andy and Kate. Moreover, the nature of everyday academic relationships that she had long cultivated in the university was crucial to the way in which the interdisciplinary project was constructed. So when Andy and Kate entered the field they already had some cultural capital to use in negotiating access, and the friendly relationships that Susie had developed with senior engineers on the grant meant that they were welcomed, generally without hostility, and often with warmth. All of this also meant that Susie was playing a distinct role in how synthetic biology was enacted in the university.

Andy was the sociology postdoc hired to work on the grant. It was his first job and it meant moving departments, cities and – to some degree – research interests. He was just finishing up his Ph.D. in Science and Technology Studies on the development of neuroscientific forms of lie detection, but he already had some significant ties to the synthetic biology community. Whilst studying for his Ph.D. he had worked with his supervisor, Paul Martin, then at the University of Nottingham, to write a report on the 'social and ethical challenges' posed by synthetic biology. The report was commissioned by a working group of the Bioscience for Society Strategy Panel of the Biotechnology and Biological Sciences Research Council and was subsequently published (Balmer and Martin, 2008). He had been chosen by Paul to work on the report because of his background in biology, which he had studied as an undergraduate student. This meant that when Andy entered the project on water innovation and synthetic biology he was able to use a certain amount of the scientific terminology and could – more or less successfully at times – make his away around a laboratory without causing a total disaster.

By this point in Andy's career the report had been disseminated quite widely, was being regularly cited by scientists, social scientists and others and was frequently referenced in national conferences, workshops and governance contexts. As such, the senior engineers in the water project were familiar with the report. Moreover, whilst it was rapidly succeeded by a vast number of similar and more substantive 'ELSI'-style publications, as well as more explicitly performative documents like the Roadmap, for a short while the 'Balmer and Martin report' was pretty much the default reference on 'social and ethical challenges' in the UK. However, it was also being taken up by scientists and engineers as the BBSRC's position on synthetic biology more broadly since they had yet to spell out a clear set of priorities for the area. Andy was thus positioned on occasion as an expert not only in how synthetic biology might have 'negative implications' but also on what synthetic biology actually was. This, as we explore in Coda 3, played a significant role in how Andy's presence in certain meetings was understood and how his questions during such participant observations sometimes caused tension.

Kate was just starting out on her Ph.D. in Science and Technology Studies. She had previously studied biology and neuroscience and developed an interest in science education along the way. She applied for a Ph.D. studentship funded by the White Rose Science Education Network on the topic of 'becoming a scientist', which had been put together by Susie in collaboration with a colleague in science education. For Susie, this Ph.D. reflected a long-standing interest in developing the sociology of epistemic communities and a background in educational research. For Kate, the theme of this studentship echoed her evolving interest in how students develop identities, affiliations and a sense of belonging within particular disciplines, specifically in relation to the biological sciences. Susie's involvement with synthetic biology at the university and the imminent iGEM competition offered fertile ground to develop questions around the training of novices in the context of an emerging technoscience. So, Kate became involved in the synthetic biology project not through any official position on the grant, but via an ethnographic entanglement with iGEM and synthetic biology as her Ph.D. field sites.

Kate's involvement in the project constituted her first tentative steps into conducting ethnographic research and she entered the field still very much as a sociology novice, concerned with her own trajectory of becoming a social scientist. It remained very much an open question what might be expected of her in the project both by her sociological and scientific colleagues. In particular, the knotty issue of what it might mean to be a 'participant observer' or 'collaborator' in this situation haunted her initial days on the project. Perhaps stemming from this early uncertainty, Kate first moved into more of an observational role with the intention of 'finding her feet.' However, she soon found that the bob and weave of everyday life on the project quickly demanded and inspired a complex range of participatory modes of engagement. Such entanglements ranged from being 'on the team' and helping with the 'human practices' dimension of the iGEM project to being seen as a fully fledged social scientist ready to advise the students on a host of sociological issues, or mobilising her own biological knowledge alongside the students in the lab. Through such shifting engagements, it soon became clear that Kate, along with Andy, had become fully integrated into the iGEM team, and so also into the project on synthetic biology and water engineering more broadly.

This was how we got started. Our own everyday and professional lives featured in how we approached the field, as well as in how we worked to collaborate with colleagues in the project. As we indicated in Chapter 1, our progress through the book is punctuated by these codas. They function as moments of reflection on various questions regarding the possibilities of and constraints on collaboration that we encountered whilst working towards post-ELSI relationships, and, ultimately, in how our efforts figured as part of our colleagues' work to enact synthetic biology in their changing practices.

2
Barriers

> Here's an invitation to the world! Can anyone solve this problem?
>
> (Water Company R&D Manager 1)

Introduction

We are in a time of grand challenges. At all levels of research policy, funding and scientific practice the impetus is to address major societal problems, using innovative approaches, drawing on the expertise of multiple disciplines. A number of shifts can be understood to underpin this challenge-led form of working, including: evolving modes of knowledge production; economics, both local and global; material transformations in the climate, Earth and food systems and the demand for accountability in the public realm. In response to this, research structures are being reorganised to facilitate delivery on these challenges. One such change regards the increasing participation of industrial partners in academic research projects, co-funding of academic work by companies and the rise of co-production of knowledge, scientific objects and new technologies as an ideal mode of implementing these partnerships. In this regard, the involvement of industry in academic work is becoming normalised in particular ways.

At the same time, as we examined in Coda One, social scientists have been enrolled into the natural sciences and engineering. The entanglement of these various actors has come about because, among other reasons, there are fears that technology will not travel from the university to the market, and that publics will rebel against innovations that might otherwise have benefited the economy through, for example, jobs and growth (SBRCG, 2012).

Our academic colleagues in the project in which we participated, and through which we sought to try out new forms of collaboration, had decided well before the creation of the grant that there would be 'barriers' to the successful 'uptake' of synthetic biology in the water industry. This was partly informed by their familiarity with these larger shifts in knowledge production and governance. However, it was also based on their own direct experience of working with water companies in the past. Some of our colleagues had decades of experience and were knowledgeable about how the water companies were organised and operated. As such, the barriers to success were seen to be multiple, inter-related and long-standing. This much was understood. What was less clear, however, was how these barriers worked exactly and whether they would be insurmountable or indestructible. Further, if they were not, the team wanted to know what exactly could be done to get over them or get rid of them. As one colleague put it:

> Synthetic biology, it is very academically interesting and exciting, but realistically there was never going to be the chance that any of this was going to be used in the water industry soon. That realisation early on meant that we were interested in finding out whether the barriers to implementation would be insurmountable, or whether they weren't insurmountable and we could be absolutely aware of them and what they were. (Academic Water Engineer 1)

As we mentioned in Chapter 1, the project, being part of a 'cross-disciplinary feasibility study', was designed to test out how well synthetic biology might do in the context of water services. Could it be used to respond to the invitation from the industry to solve their problems? In order to find out, the project sought to address several barriers. These included the water industry's 'conservative attitude' to innovation, which was understood to be a potential limit to interest in the high-technology of synthetic biology; the 'over-regulation' of the industry, which meant that their hands were tied even if they were interested, and the 'public's ignorance' of water treatment services, which meant that the companies could not charge consumers more for water and so could not afford to pay for high-tech research anyway. Finally, and although it was not made explicit in the final version of the grant application, some of our engineering colleagues also saw the possible 'public fear of synthetic biology' as being a barrier to getting the water industry to accept the field because it was cautious about what consumers might think. In this chapter we explore how such anticipated 'barriers' to the successful implementation of synthetic biology in the water industry were conceptualised, articulated and enacted through the everyday practices of actors in the project.

In this regard, we are concerned here with how our colleagues tried to make synthetic biology into something that was palatable to the industry and shape the kinds of value that the field could have in this situation. We are also thus engaged in trying to understand how different promises and practices were fitted together, or not, as attempts to surmount these barriers were made. As Frow (2013: 444) argues:

> The registers of epistemic, academic, ethical and market value we encounter [in synthetic biology] are centred at different levels of abstraction, operate on different timescales and are concerned with different objectives. This is not [about] a simple gap between the promises and practices of synthetic biology; it shows how value is being constituted in these in-between spaces, how it requires both promise and practice to emerge and form, and how it both shapes and becomes embedded in the material and social worlds of an emerging field.

In our account, we examine this very issue, extending Frow's argument by considering the practical enactment of synthetic biology within a specific industrial situation and through the everyday features of academic life. By doing so we also articulate how responsibilities for bringing synthetic biology into being are distributed such that the onus to change practices and overcome 'barriers to innovation' rests squarely on the shoulders of academics, whereas the capacity to bring about such changes is far from within their control.

First, we explore how synthetic biology has been constructed at the core as an interdisciplinary field designed to service industry. Second, we show that the water industry's focus is instead on consumers and stakeholders. The approach to innovation in the water industry is significantly shaped by the contrivance of markets in water services provision, which are a product of privatisation in the 1980s and of current public health regulations. We then examine how our colleagues worked to synchronise these two worlds, detailing how they used the promissory narratives of synthetic biology to link their high-tech science with the concerns of water companies through 'global challenges', how they tried to get industrial actors excited about the field and how they struggled to synchronise academic and industrial practices to bring about the vision of synthetic biology in the local situation.

An interdisciplinary science in the service of industry

In 2011, a meeting was convened by the Technology Strategy Board, part of the UK government Department for Business, Innovation and Skills (DBIS), under the auspices of the Rt Hon. David Willetts MP (then Minister of State for Universities and Science) and the Rt Hon. Dr Vince Cable MP (Secretary of State for Business, Innovation and Skills). As the briefing materials described, the round-table meeting was intended to bring together academics from a range of disciplines alongside industry experts and governance actors to discuss: 'What makes [or] will make the UK an attractive place for Synthetic Biology?' and 'What are the potential barriers?'

As we have so far described, synthetic biology is very much about making biology easier to engineer in order to turn biotechnology into a more robust and predictable industry. As such, the field itself is explicitly concerned with overcoming barriers to market. For example, SynbiCITE was funded in the UK as a centre designed 'to provide a bridge between academia and industry to speed up the development of new technologies in synthetic biology' (Smith, 2013b). Launching the centre, one of the core figures in synthetic biology in the UK, Professor Richard Kitney, claimed that:

> One of the major challenges that industry and academia face in synthetic biology is translating breakthroughs in research into new products. The aim of the new Centre is to break down road blocks so that new industries can be developed, which could ultimately help to safeguard the UK's economic future.

The emphasis on industrialisation in synthetic biology emerges from frustrations with the speed, fallibility and unpredictability of existing practices for making novel microorganisms and turning them into economically viable, commercially relevant products. However, it is also situated in a longer history, not only of the confluence of engineering and biology (Campos, 2012) but also of the reconfiguration of scientific practices, governance and capitalism, around the construction of biovalue (Vermeulen *et al.*, 2012) for the future bioeconomy.

How science should work and how it should be funded, made accountable, used for social goods and so forth are questions answered increasingly through its potential to contribute to the growth of the national economy, and so to ensure that the UK remains commercially competitive on a global scale. However, there is still a concern that academia and industry are too far apart. Much work thus continues to go into restructuring the relations between knowledge production and innovation. The 'divide' or 'gap' between academia and industry is widely associated with a problem of industrial and commercial translation, and feeds the mantra that the UK can invent, but fails to capitalise on its inventions (Moran, 2007). Indeed, this divide and the associated failures to industrialise academic discoveries have become known as the 'valley of death' (HCSTC, 2013) into which promising innovations tumble, never to see the promised market. In order to string a bridge across the 'valley', multiple schemes and structures have been devised in academic institutions, government departments and the funding councils, as well as in corporations. For example, funding devices have been created and new organisations have been set up by the government, like the 'Knowledge Transfer Networks' and 'Innovate UK', with the explicit mission of identifying and investing in sectors seen to have the potential to bolster economic growth by better bridging the gap between academia and industry.

Engineering has been especially central to these ambitions. This is at least partly because the very point of engineering, for most academic engineers and certainly for our colleagues in the project, is to develop useful applications that fix a problem, particularly with respect to industry and the civic infrastructure (Bucciarelli, 1994; Vincenti, 1990; Vinck, 2003). Engineering is thus highly instrumentalist and often geared towards economic goals, but nonetheless is understood by engineers to ultimately be in the service of the people and civil society. As such, the emergence of synthetic biology – with its emphasis on the use of engineering practices to industrialise the production of microorganisms to service industry – has to be understood against this long-standing concern with bringing about the promised knowledge economy. That the service of engineering to society is increasingly articulated through economic and industrial imperatives is particularly acute in the field of synthetic biology.

In addition, synthetic biology's economic promise is tied to its potential to ameliorate a number of more or less global or 'grand' challenges. Such challenges include climate change, the energy crisis, food and water shortages, pandemics and so on. Grand challenges like these have long been understood to be too complex for any single discipline or industrial sector to overcome and it is only in recent years that significant efforts have been made at national and international levels to shape innovation processes around them (Cagnin *et al.*, 2012). In this regard, the solution of grand challenges is also understood to be through market mechanisms and technical innovation.

Synthetic biology has emerged at the confluence of the various socio-economic movements described above. As such, the field, or at least its rhetoric and promises, has been designed in light of these shifts, and so it looks rather like an archetype, tailor-made for the contemporary era. On this note, Robert Carlson (2007), a prominent spokesperson for synthetic biology, has nicknamed our present period the 'paleobiotic', drawing a comparison to earlier human periods of technological development, and thereby suggesting that our present capacities in biotechnology are the equivalent of the Old Stone Age. Pipetting by hand, using PCR machines and designing bespoke microbes for particular problems: these practices, we are encouraged to think, are the equivalents of hammerstones and hand axes. Synthetic biology, he argues, is one way in which we might move into the 'neobiotic' period, which – it is implied – will be characterised by the same kinds of socioeconomic transformations that occurred when humans began farming and using more sophisticated stone tools. As we described in Chapter 1, the more common temporal comparison made about the field is that it represents the beginnings of a second industrial revolution. Either way, such analogies draw our attention to how synthetic biology's adoption of engineering practices, with its close ties to industrial instrumentalism, is intended to bring about major shifts in how civil society manages its problems, or, as has been claimed, to transform how we 'heal, feed and fuel' ourselves (DBIS, 2013).

Emphasising how shifts in epistemology, technology and knowledge production practices will bring about epochal change in the socioeconomic ordering of everyday life places most of the blame for the failures of biotechnology on the shoulders of academics. Indeed, the onus so far has very much been on academic scientists and engineers to change their practices to align better with those of industry. Contrary to the wasted energies of past biotechnological efforts, designing bugs on a case-by-case basis, synthetic biologists promise to bring about innovation in a way that fits directly into industrial processes. Meeting the demand for control, prediction, large-scale manufacture and low-cost production was, on paper, exactly what was required for the challenges posed in the water industry. In this regard, changing their practices around the emerging norms in synthetic biology seemed like an ideal fit to our colleagues. An emphasis on economics-driven science, so that basic research at the bench was done with a commercially viable outcome in mind, seemed exactly like the kind of thing they needed in order to persuade the water companies to get interested in more high-technology solutions to their problems.

The emphasis placed on interdisciplinarity in synthetic biology also greatly appealed to the engineers in the project team. Indeed, the RCUK's investment in the networks in synthetic biology, which preceded our project and one of which was funded at the university, were explicitly designed to bring together actors from different disciplines, precisely because synthetic biology was understood to be fundamentally inter-disciplinary. As an early report on the field from the Royal Academy of Engineering (RAE) explained:

> These networks should help to overcome some of the difficulties of developing such a new and interdisciplinary subject. They will allow researchers to overcome language and terminology barriers, establish productive partnerships and stimulate ideas – thereby establishing an effective critical mass of researchers in synthetic biology from across all the necessary disciplines (RAE, 2009).

The project leader saw it as one of her roles, not only in this project but also at the university more broadly, to bring together academics from different disciplines to study problems of mutual interest and so to be better able to solve those problems. This was part of why synthetic biology appealed to her, as she said 'I like the idea of the combination of all of the different people' (Academic Environmental Engineer 1).

More generally, working across disciplines is often taken to be a silver bullet for approaching and solving tricky problems and grand challenges. Inherent here is the view that academic disciplines have them-selves created barriers, not only between themselves, but also between knowledge production and reality. Such barriers are seen to disable productive work (Barry *et al.*, 2008). And, as Kearnes and Wienroth (2011: 47) explain, there is thus an expectation that 'inter- and multi-disciplinary research will likely produce more complex and successful approaches and solutions to societal challenges.'

So it is no surprise that the mixing of disciplines is a foundational brick in the edifice of synthetic biology and that it is neatly cemented next to global challenges:

> Multidisciplinary expertise is already enabling the UK to make signifi-cant contributions to international research programmes, and also to assimilate and respond to global developments as they arise (SBRCG, 2012: 4).

This approach to synthetic biology as an interdisciplinary science, designed to service industry, in order to solve grand challenges was very much adopted in the construction of our project, as was written in the grant application:

> Success in this new area [of water challenges] requires communication, cohesion and collaboration from engineers and scientists within an appropriate ethical, social and legal framework.

Such cross-disciplinary collaboration was not new to our project participants. The engineers had been trying to solve water problems by working across disciplinary boundaries for some time; indeed, it had been their struggles to solve problems from within their own disciplines that had brought a number of them together.

> I've had a long standing thread of research on discolouration, which is probably the biggest customer complaint you get. It accounts for about a third of customer phone calls that water companies get around the world. I've come at it from pure engineering, physical processes. You can get so far with explaining and predicting what's going on, but I realised that microbiology and chemistry were having a pretty important role. So I started working with [Academic Environmental Engineer 1] and through that started working with [Academic Microbiologist]. (Academic Water Engineer 1)

Overall, the core emphasis in synthetic biology on industrialisation and socioeconomics, global challenges, engineering and interdisciplinarity sat well with our colleagues' existing practices and with their concern to improve innovation in the water industry. Indeed, it was the terminology of synthetic biology, with its use of the engineering lexicon, as much as the practices that appealed to our colleagues. Using it, they articulated opportunities to get engineers and molecular biologists to work together more closely.

> You don't get engineers as excited about molecular biology as you have done with synthetic biology. And maybe molecular biologists are right, that it is the same DNA tools, techniques. But you're playing with engineers now, when you wouldn't have done if it wasn't called synthetic biology. (Academic Environmental Engineer 1)

The shift in terminology and the associated constitution of a novel field was part of how our colleagues hoped that they could make their research 'stick' in the water industry. In this regard, synthetic biology offered a *sticky* terminology (Molyneux-Hodgson and Meyer, 2009) that helped to consolidate a particular way of doing interdisciplinarity among all the participants, entangling individuals, companies, problems and materials into an evolving assemblage.

However, as we have already indicated, they knew that this was still going to be a hard sell. Our colleagues knew from experience that even when there is reasonable agreement between academics and industry on specific industrial needs, the two worlds can remain far apart, and not for want of trying to climb the barriers. Our colleagues actually described the example on which they were working, the development of a synthetic biology biosensor to detect pathogens in drinking water, as being a possible case in point. Indeed, as our own sociological investigations and interactions with the industry developed, it became clear that for the biosensor to be used by the water companies, the following would be required: re-skilled workers, designed ease of use for field workers, a fast detection-response time, robust use in field sites, sensible detection limits, adherence to water quality regulations and a speedy and determinate reporting function. Even if all these criteria were met, and to all intents and purposes the sensor functioned, this would not guarantee implementation. This was because, in the view of both academics and industry actors, the water sector is unadventurous and orthodox. In describing their ongoing work on the biosensor one colleague still worried that:

> If we could get the technical issues to work there is absolutely no guarantee that anybody in the water industry would ever be able to progress with the ideas. The water industry is quite conservative. (Academic Water Engineer 2)

Indeed, from the academics' viewpoint, the water industry could thus be conceptualised as a testing ground for synthetic biology's broader chances at solving interdisciplinary, economically relevant, global and grand challenges. The perceived conservatism of the industry, its scarcely established connection to high-tech industrial biotechnology, the context of regulation and potential for 'public fears' represented to our colleagues a kind of hard case in assessing the potential of the field:

> So [we asked] in an area that, on the face of it, might be quite tricky to apply some of these concepts, can they apply there? At every level, from would the customers, would they accept it, through to the regulators. Because there would be different problems in this area than

there would be in pharmaceuticals or biofuels or whatever else. So for me, it was [a question of] could some of the same tools we've used in that area [of human tissue and bio-engineering], be imported over into an area that would seem to be quite, maybe more divorced from what you would see as a core area for synthetic biology. (Academic Chemical Engineer 1)

Synthetic biology became a test case for the water industry, regarding whether they would or could accept high-tech solutions, and the water industry became a test case for synthetic biology, since if they could get it to work in this barrier-laden situation, then engineering practices would clearly have won the day, and the valley of death might just be crossable. However, much of this hung on what exactly our colleagues meant when they said that the industry was 'conservative', and so how it would be likely to react to the academics' innovations.

A conservative industry in the service of consumers and shareholders

There was no shortage of ideas on what hindered innovation when this was discussed with project colleagues. Interestingly, most of these problems were seen by academic and industry actors alike to reside primarily within the industry camp. A report produced by the project team in the first few months generated a long list of barriers that cut across financial, organisational and cultural concerns. Some of the issues relating to innovation failures in the past were seen as signals of long-term trends found in the UK industrial base at large, such as reduced attention on research matters in the privatised world:

> R&D groups are generally downsizing, there is less available manpower to trial and test innovation, and innovation is now mainly conducted by the supply chain or by international technology developers. (Internal Project Report: 3)

Moreover, fractured supply chains and uncertainty in future needs led to reduced investment in both research and development overall:

> The supply chain is diverse and fragmented, consisting of small companies operating on low margins, low sales volumes and consequently small development budgets. Small supply companies have little opportunity to trial and prove technologies in real world

environments. When regulatory approvals are required, the costs can be prohibitive. (Internal Project Report: 4)

In this regard, there was a significant amount of knowledge in the project already about why the industry invested very little in high-technology solutions to long-standing, intransigent problems. However, this under-standing of innovation trouble was not laid only at the feet of actors in the water companies. Instead, actors in the project felt that each barrier to getting engineering solutions into the water industry was connected, in one way or another, to how the companies were regulated.

Following the privatisation of water services in England and Wales in 1989, the size and complexity of the infrastructure meant that compa-nies were set up regionally, each taking responsibility for their local water sources, supply and drainage systems. As such, consumers have no choice in their water supplier. This means that there is very little opportunity for competition within the market, leading to a so-called 'natural monopoly'. One anticipated consequence of setting up the market in this way was that it might result in poor service and increased costs to consumers. As such, a regulator, Ofwat, was created at the point of privatisation to oversee the industry in order to prevent these prob-lems. Ofwat is responsible for setting price limits and reviewing these limits every five years. It aims to strike a balance between the cost of the service to customers and the other responsibilities of the industry. These other responsibilities include providing a return to investors, making improvements wanted by customers, ensuring maintenance of existing pipes, sewers and treatment works, meeting drinking water standards and environmental obligations, and balancing supply and demand.

Water systems in the UK are further regulated by several other organi-sations, including the Environment Agency (EA), and the Drinking Water Inspectorate (DWI). These multiple layers, with differing remits, create conditions that are easily described as 'heavily regulated'. However, the regulation has developed in this way not only through responses to neoliberalism and privatisation, but also because of the material fact that water is a requirement for life. As such, regulatory structures must protect humans and other organisms from potential hazards in drinking water and sewage, and must thus also protect water sources, ensuring that they are maintained and free from contamination. In this regard, the multiple layers of regulation make it is easy for water company actors to argue that regulation works against the use of radically new technologies that might provide step changes in performance or in cost. Their attitude to inno-vation is thus, they sometimes claim, determined by the importance of water for public health. Indeed, this message was sometimes put forward by R&D managers in the water companies. In the following quotation, for example, an R&D manager blames the regulators for being conservative, and suggests that this does not necessarily reflect the nature of risks in the industry, but rather their (public) perception:

> I think our regulator is cautious ... and sometimes regulations can be related on perception, as well as actual risk. (Water Company R&D Manager 2)

At the same time, R&D managers themselves sometimes understood the citation of regulations to be a bit of an excuse for inaction:

> if there's a compelling case to do something, then it will get done ... I don't really hold with the fact that regulation gets in the way. (Water Company R&D Manger 3)

And some suggested that the lack of uptake of new technologies could often be down to other factors:

> Many managers prefer to have the same technologies that they are comfortable with so they can simplify the stock maintenance, training, and so on. It also gives them a knowledge powerbase over the 'young whippersnappers' with their new-fangled gizmos. (Industry Broker 1)

So it is unclear whether the industry is actually chomping at the bit to innovate and that, if unshackled from the regulations that protect the environment and consumers, it would leap forward into high-technology research given the chance. Indeed, there are several contexts in which the water companies continue to use existing and traditional approaches long after new, more capable technologies have become available. This is primarily because the investment required in the infrastructure is weighed against other concerns, such as lower returns to investors, the resistance to change from actors within the companies and whether or not there is any regulation forcing companies to improve services in that particular direction. For example, as one R&D manager described:

> What's quite common with water-related R&D is that you get to the stage where the technology works in the lab, and because you haven't

involved the supply chain at the outset and understood all the issues
of getting the technology to the market, you get to a position where
the costs of tooling up and marketing the technology are just so
prohibitive. You get to a point where the commercial organisation
doesn't recognise, or aren't willing to accept, that the sales volumes
are going to be high enough to justify that initial investment. (Water
Company R&D manager 4)

This argument principally concerns the ways in which the water industry
organises its expenditure on capital, R&D and consumer relations. It
is about ensuring that the business is profitable and makes money for
shareholders. Indeed, this particular 'valley of death' has been shaped in
particular ways by the privatisation of the industry. In constituting the
water companies as natural monopolies, the Thatcher government effectively removed any imperative to invest in R&D. Before privatisation,
investment decisions were taken by water companies independently
of one another and knowledge flow between companies was limited:
'something might be going on in Yorkshire and there'd be no knowledge [of it] in Severn Trent' (Water Company R&D Manager 3). Indeed,
research and development was one area in which the then public utilities competed with each other. Nowadays, communications between
companies have improved and various knowledge management systems
are in place. Shifts in research funding have also occurred, so that
public funding is far more forthcoming and organised around industrial
concerns. As such, most research on water engineering is conducted by
universities, and so funded by public taxes, not by the companies that
profit from those advances. It also means that the companies are more
likely to work together with universities, so that they can support bids to
research councils and all benefit from technology outputs. The natural
monopoly provides incentives to such collaboration since there is no
risk of losing consumers to other companies.

Nonetheless, for academics, the regulation of water provision and
sewerage systems, designed to protect consumers, is at the heart of
why the industry is so conservative, and so of why there might be a
barrier to synthetic biology's entry into the water market. This was a
constraint to which our colleagues returned again and again through
their work. In meetings, labs, interviews, 'industry days', grant applications, papers and so on, the conservative nature of the industry as
regards technical innovation was understood to be caused by regulatory
structures. For example, our colleagues regularly referred to the example
of pipe leakage, which is one of the ways in which several critical global

problems, including energy, climate change, water scarcity and so forth, manifest locally for water companies. Academics felt that they had a number of possible solutions to detecting leaks in pipes, but that the conservativism of the industry prevented them from successfully getting these innovations across the valley of death.

> Academic Water Engineer 1: You only have to look at the problems and challenges the industry has to see the importance of critiques in terms of the water industry taking on new technologies. [Innovation in the industry has a] relatively incremental nature, so like a new way of detecting leakage with noise, takes ten years to get to industry. It's a tried and tested technology, they already use it, it's just developed from the same technology, but it takes them ten years to do it. [In this project] we're talking about a total re-think, a total change of what they're going to do. There's a huge barrier there, from every-thing, from the kind of the pure company perspective: What's the cost-benefit? What's the confidence of it? Through to the Drinking Water Inspectorate, and the high-level of policy, of the government saying it's okay. Also of the public perception of accepting what we're trying to do. [...]

> Andy Balmer: And why did that take so long?

> AWE 1: Just because the industry is so conservative, so highly regu-lated. And because of Ofwat they're pinned down so much in terms of the benefits they can make, the improvements they can make. They're very driven to do what the regulator says, so they're conserv-ative. They're trying to do what appears to be a sensible cost-benefit for the customers.

Regulation and consumers were thus the common targets of academic complaint. It was extremely unusual for them to draw attention to the fact that the water companies are precisely that, companies, whose primary purpose is to provide water services in order to make a profit. Shareholders, profit mechanisms and the natural monopoly were rarely attributed any blame for the organisation of industry R&D. Instead, overregulation was blamed, and so were ignorant consumers.

This complaint of overregulation was also the subject of The Cave Review (2009), which was commissioned to examine the state of the water industry after 20 years of private operation. The review exam-ined competition in the water markets and made recommendations for changes to the regulatory and legislative frameworks of the sector, in the interests of encouraging innovation through competition or cooperation. Two key sociotechnical challenges were identified in the report, namely, population growth and climate change. Both of these are anticipated to lead to increased demand and lower supply, and would create demands on the industry to find better ways of allocating, treating and using water. Not only will the industry have to meet the needs of users but it will also need to improve environmental outcomes and reduce its own environmental impact. At the same time, reports on water regulation almost always highlight the need to ensure a fair price for water consumers, given the absolutely essential nature of clean drinking water and an effective sewerage system for public health. The recommendations of the review were thus aimed at increased competition and innovation within the industry as means to address these global challenges, improve services to users, increase efficiency and so also to reduce operating expenditure and keep prices down. Ofwat agreed:

> Effective competition has the potential to bring benefits in a range of water and sewerage services, [and] is a key driver of efficiency and innovation. (Ofwat, 2015)

Water industry actors were less convinced that innovation could be spurred forward by changes to market emulation mechanisms. The basic fact to some of them seemed to be that the market for water simply could not operate like other markets apparently do. For example, regulators and academics often compare the water industry unfavourably to the pharmaceutical industry, citing the latter as a model for innovation. However, for R&D managers in the water industry this was not a relevant comparison, and failed to recognise how innovation practices were incentivised within the business models on which water companies operated:

> The Cave Review is about trying to get more competition into the water industry. In regulated industries, especially if it's like the water industry, it's quite difficult to generate [competition] between the companies, and I think there's a view from the regulator that industries that are more innovative are more competitive. Like pharmaceutical industries, if they don't innovate they go out of business. Very live or die. I don't think the water industry is like that. The investment in research is much lower and the level of growth of the industry is much lower, we're not set up as growth businesses in the same way as a company that has got access to reward markets. (Water Company R&D Manager 1)

Nonetheless, one way in which Cave's recommendations have been implemented is in the development of the 'service incentive mechanism' (SIM). The SIM is a device designed to encourage the water companies to provide better services to customers whilst also improving innovation. The SIM essentially allows for comparison between different water companies and financially incentivises improvements on the basis of where companies fall in those comparisons. This is intended to bring about a form of competition that tries to mimic market conditions. 'Service' is measured using quantitative and qualitative indicators. Quantitatively, the SIM measures the number of customer contacts or complaints to a water company when there has been a problem and determines various other scores via customer satisfaction surveys. The qualitative component involves a survey of 800 customers over the course of a year. The two components are put together to give the annual SIM score for the company. The SIM then acts as a financial incentive for improving service because it impacts on the prices that the company is allowed to charge their consumers for water services. For example, the SIM scores from 2011–12, 2012–13, and 2013–14 were used to set the price limit adjustments for water companies in 2015. Companies with a better score are rewarded with higher price limits, and those with poor scores are penalised.

In this regard, innovation systems, profit mechanisms and regulatory structures are very much entangled in the water industry, through the contrivance of markets (MacKenzie, 2006; MacKenzie, 2008; Molyneux-Hodgson and Balmer, 2014). The idea of this new governance form is to generate such reward markets artificially so as to mimic existing, high-tech, high-innovation companies, particularly those in the idealised pharmaceutical industry. However, we know that novel techniques and technologies in the pharmaceutical sector have not produced a revolutionary translation of research into applications (Nightingale and Martin, 2004) and the realisation of new products has fallen far short of expectations. The faltering future, for example, of a personalised kind of medicine promised by pharmaceutical innovation is a good mirror to the promise of catastrophe aversion being embedded in water innovation and regulation.

The privatisation of the industry and the regional set-up of the water companies, along with the essential nature of water and sewage management, mean that regulators have to mimic market mechanisms to spur innovation whilst restraining risky business practices to protect consumers from price hikes and to ensure that the companies both meet water quality standards and try to tackle global problems.

These changes to the governance of the water industry had come about just prior to the project on synthetic biology being designed, but had long been in the making. The regulation of the industry and the creation of supposedly market-like mechanisms seemed to bode well for the arrival of a high-tech field like synthetic biology. All our colleagues had to do was to get the water companies interested. One way in which they did so was to try to sync narratives about global challenges in academia with those in the water industry.

Syncing narratives

As we described earlier, knowledge-making practices have increasingly been organised around global challenges. Similar shifts have also been afoot in the water industry. Problems that were once construed as local, for example pipe leakage and aging infrastructure, are now interpreted as part of global challenges. In this regard, intractable global problems are understood to manifest locally. This 'glocalisation' (Swyngedouw, 1997) of water industry problems is something that has been brought about through a range of shifts in materiality, infrastructure, economics, politics and regulation (Biro, 2007; Swyngedouw, 2005). Our colleagues in the project certainly saw this as a distinct opportunity to connect the discourses in water services provision up with those in synthetic biology.

At the same time as connecting the water industry to synthetic biology through global challenges, the engineers in the project also wanted to manage expectations regarding the timeframe in which the field could contribute to solving such massive and complex industrial problems.

> Andy Balmer: So what do you think the water industry will get out of this project?
>
> Academic Water Engineer 2: A better awareness. They're looking for what's going to help them, not in this five year period, but in the next five year period. Typically when you work with industry you've got two things, a short term deliverable, that takes twelve months and can be implemented immediately, then something that will give them a strategic benefit over a three–five year period. This project has potential not for this period but for the next one, so to some extent this is a loss leader for the industrialists, but they can see potential in it, not immediate potential, but potential a few years down the line.
>
> AB: And what is the potential? The potential to do what?

AWE 2: The big challenges they face at the moment are energy use, potential pollution of the environment and the fact that the climate is changing. And they don't have too much money, so the climate is changing *and* their infrastructure is degrading. The two drivers are going in the opposite directions. So they're looking at ways to reduce energy – that's a big one – and they can see that being a long-term issue as well. [They also want] ways to improve or monitor their infrastructure, so they can maintain a level of service even if the climate is changing and the infrastructure degrades significantly. Those are the major challenges and I think that synthetic biology has the potential to help in the long-term of each of those cases.

Making links between the glocalisation of water industry problems and promissory narratives in synthetic biology meant that the field was being pitched, in the local situation at least, as a long-term investment. Indeed, by the end of the project, when it came time to apply for more funding to work on water industry problems, the same issues were still at stake. A follow-up grant linked governance pressures, material failures and global challenges thus:

Against a background of increasing expectations, ageing infrastructure and regulation, urban water systems are under escalating pressure to deliver safe and secure water supplies, hygienic sanitation and sustainable drainage whilst facing global challenges of climate change, flooding, drought, carbon targets, demographic change, urbanisation and cost. (Quotation from a follow-up proposal)

Of course, no one had ever been under the impression that synthetic biology, in less than two years, would enable the academics to solve climate change or even pipe leakage. Rather, their aim was to encourage the industry to become more speculative, to 'think big' about biology, and to begin to take greater risks in technological innovation. A critical route into this was to link synthetic biology to the water industry through how timeframes were being reconfigured in the water companies via regulation. In this way, our colleagues positioned synthetic biology as a field that would allow the industry to make use of the emerging changes in the regulatory organisation of the water market.

Contrary to the more traditionally prolonged nature of academic work, industrial temporality is created by the regulatory system, which requires planning reports every five years and rewards company performance and efficiency on the basis of a five–to–seven-and-a-half year benefit.

Annual cycles of profit also impact on the opportunity for innovation. The significance of such timescales for innovation was outlined in the Cave review:

> The current framework of economic regulation does not always encourage significant investment in research and development or the trialling or adoption of innovations. This is because, depending on the level of investment, the probability of a successful outcome and value of the saving, the current outperformance period of between five and seven and a half years may be insufficient. ... While such a system ensures that companies consider the short-term value-for-money of spending, in the long-term it may also lead to a decline in research and development and in companies' ability to drive innovation, which will be important in meeting the new challenges of climate change and population growth. The short-term protection of customers may therefore come at the expense of long-term industry performance. (Cave, 2009: 17)

What Cave proposed is a change to the regulation of time in the industry to shift the practices of innovation. The academics in the project tried to encourage the water company actors to think about synthetic biology as a good case in which to implement some of the review's proposed changes. Following some encounters in which project actors met with industrialists, one R&D manager responded:

> I'm looking more at five, ten and maybe 25 years for this [synthetic biology] to be fully developed so I've got no short term expectations. (Water Company R&D Manager 1)

So, they were definitely hearing that academics wanted this to be a long-term goal. Ensuring R&D managers did not have immediate expectations for academic outputs was very much about negotiating a change in timing. Normally the water companies expect academic partners to produce industry-relevant outputs within their five year investment cycle. In this regard, pitching the integration of synthetic biology into water R&D as a longer-term goal was sensible. At the same time, however, the academics knew that they would have to do something quickly because they also had to satisfy the funding body and 'make something groovy'. In this regard, the funding regimes for academic science hastened up the academic's research whilst they were trying to slow down the industry.

This is in large part because funding regimes of universities are under pressure from the current UK government, which academics tend to think demands far more in terms of outputs in short timeframes than previously required. RCUK's initial funding in synthetic biology, for example, was quite short term, and it is only in recent years that longer-term projects (of five years) have been funded. Working on a shorter-term project, with long-term goals, hoping to convince the industry but needing more immediately to satisfy the RCUK was a delicate temporal balancing act. As one colleague explained:

Traditionally, academics have more time to think about things. Whereas if you've got a group of investors at a board meeting and you're trying to produce intellectual properties or patents, and you've only got six months' money, however much you might like the luxury of spending five years on it, you're just going to run out of money and no one's going to be paid. I mean it's a trade-off. If you have to do something fast, you have to take some gamble and you can't do all the back-up experiments to make šure that you're going to neces-sarily get it right first time.... Like I say, I think it's mostly to do with timescales and what money is available. And the thing is, I think as more government money is cut, as the EPSRC is cut, more and more people are going to find themselves in the situation that I'm in [in this project], where you have to hit the floor running, make a deci-sion and try it for a year and then get out and hope for the best. That's not, I mean that's quite a hard thing to do I'd say.... The universities can rise to the challenge but it's a different way of working than one is used to and whether you're losing something in the process in terms of finding out things by serendipity, having a bit more time to think about things so you don't waste time getting them wrong. I mean thinking is man hours. Time will tell. (Academic Molecular Biologist 1)

Cave argued that extending the planning and innovation timeline to 25 years and changing the reward system for efficiency and service provision would lift the barrier to innovation. This, it is argued, will help water companies to address global challenges by investing in research. However, it was understood that this might pose a challenge regarding how best to protect consumers from price hikes, which are known to be significant factors in contributing to 'utility poverty'. Moreover, the review largely assumes that companies will immediately implement bolder innovation programmes to drive down their costs, increase performance and rake in extra financial rewards and consumer profits. As we've seen, however, water company R&D managers are less certain about this, as they point towards a more complex series of regulatory and everyday practices that constrain innovation or make actors in the industry resistant to change.

Whilst the Cave-proposed innovation, temporal space was welcomed by industry and academics on paper, it was also seen as creating new problems. Long-term plans were, for some, not particularly useful, because of the changing environment and financial climate 'it is very hard to plan 25 years in advance for R&D [...] it's very unlikely you're planning more than five years out' (Water Company R&D Manager 2).

In this regard, the introduction of a long-term planning requirement in the water industry did not immediately bring about a space conducive to investment or interest in synthetic biology. Certainly, our academic colleagues knew not to expect the industry suddenly to invest in the field. Indeed, their immediate drivers were coming from the academic side of things, and producing 'something groovy' for the RCUK did not necessarily mean something workable for the industry, as we explore later.

Instead, they hoped that they could use the new planning requirements to 'put synthetic biology on their radar' (Academic Chemical Engineer 1) and so to begin to open up thinking, rather than their bank accounts. What our colleagues wanted to achieve, then, was to synchronise the timeframes of the industry and their academic work and to get water companies interested in synthetic biology but without any immediate expectations of products or solutions. They did this by taking advantage of emerging shifts in the regulation of time regimes in the water market.

There was a certain degree of serendipity in this. They were in the right place at the right time, and they had connections to synthetic biology and to the water industry. However, there was also an underlying structural shift that made it possible to synchronise the narratives in this way. The emergence of global challenges into governance practices has meant that efforts are being made in a number of contexts to reconfigure practices so as to respond to such existential threats. These changes have made synthetic biology a sticky terminology more broadly, as it has emerged in the UK and the USA and in our situation, specifically, for pitching it to the water industry. In addition, the language of synthetic biology, specifically the emphasis on engineering terms, helps stick together actors from different engineering situations in order to try to respond to such global challenges. Using terms like 'parts', 'devices', 'abstraction' and so forth was handy, in that R&D managers from water companies could also speak in this way. It helped to make biology relevant to industrial problems like leaky pipes and sensible to engineers unfamiliar with the language of genetic engineering. This helped to stick together the academic and industrial actors and, on the face of it, was helping to get high-technology science across the valley of death.

However, our colleagues did not only want to get biologists, engineers and industrialists talking the same language. Our colleagues also wanted to begin to reconfigure the vision of the future for the water industry, to test out the bounds of synthetic microorganisms in its infrastructure, and – ultimately – to explore the potential for shifting the practices through which innovations in water were co-produced at the nexus of industry and academia. This would involve more than just talking with new terms and saying things about global challenges. Whilst this helped to get R&D managers interested, it had to be more than just talk.

Syncing affects

Our colleagues wanted water company actors to get excited about synthetic biology. After all, academic engineers were excited about it. If they could communicate why synthetic biology was exciting they hoped they could make a step towards getting the water companies to commit more materially and politically to the project of integrating high-technology research into industrial R&D practices. Through synthetic biology's promise, they hoped to bring about the Cave Review's imagined future. How to take advantage of the new conditions for cross-boundary working, and the shifting regulatory landscape, was the focus of much debate in the design and enactment of the project. It was hoped that these changes might just make for the right conditions for synthetic biology to flourish, so that the engineers might consolidate the shifts that were occurring in their own academic practices, as regards the adoption of synthetic biology norms around standardisation, sharing and so forth.

In the end, the project adopted a set of 'industry days', held at specific points in time through the course of the project, and organised as a key device for interaction between academics and industry managers. The initial industry day was designed as an open discussion of the major challenges facing the industry and as a dialogue around 'wild ideas' to address those challenges. Following on from this, a shortlist of ideas generated by the industry day was used to think through innovations in more depth and to explore ideas for the iGEM project and other 'proof of concept' lab work.

At the first industry day, an academic colleague gave a standard account of 'What is synthetic biology?' Following the same kinds of tropes we've explored so far, they also described some of the challenges in the field in both technical and social terms. We then held discussions on where synthetic biology could be applied in the water industry. This was a 'brainstorming session on the drivers, challenges, barriers and applications' that might be related to integrating synthetic biology into water industry R&D. We were encouraged not to hold back or think about constraints. This certainly had an effect. At one point, a particular discussion group of water R&D managers and academics envisaged creating biologically active swimsuits.

So a long list of areas for potential innovation was quickly assembled: sewer unblocking; nutrient removal; stopping leaks through self-healing pipes; energy use reduction; hazardous substance treatment; pathogen detection; designer water for improving health; discoloration; destruction of fats, oils and greases and so on. When speculative ideas were mooted and a 'Why not?' was posed, small group discussions thrashed out the issues face-to-face. Academic ideas were proposed, talked through and faced varying degrees of interest and rejection from the industry representatives. Water company managers brought up their challenges and their constraints. Through the messy process of talk, laughter, debate, argument and note taking, the groups engaged in the sorting of interests and took up positions in relation to the proposed ideas. 'Doable problems' (Fujimura, 1987) were acknowledged and fun ideas kept us all going throughout the long day.

The water company actors were quite caught up with synthetic biology and found it very interesting. A number of them were scientists or engineers by training, and a couple had been biologists at some point in their careers. They were animated by the field because it was far more experimental and adventurous than the mundane R&D issues with which they dealt on a daily basis. Getting bacteria to do wild things was far more exciting to talk about than incremental tweaks made to models of pipe flow or the slight improvement of a pump.

At the same time, the industry actors brought up significant concerns regarding the use of bacteria in the clean water infrastructure. They felt that genetically engineered bacteria would not be appropriate for use in this context since the whole clean water system is organised around keeping bacteria out of the pipes. This issue is something that we return to in Chapter 4. For now, it is enough to say that existing practices in the organisation of water services made for immediate constraints on the kinds of innovations that could be pursued if they were ever going to be relevant to water companies. The water R&D managers also drew on their knowledge of the regulatory agencies, the water consumer representative group (the Consumer Council for Water, or CCW) and of their consumers. They argued that any innovations using genetically engineered bacteria might result in controversy. For the regulators this would have to do with whether there might be risks involved for public health; for the CCW it would be about whether the industry passed on the costs of such high-tech innovations to the consumers and for the consumers it would be about both.

As we investigate in more detail in Chapter 4, there are long-standing politics in the arrangement of water services when it comes to public health and the distinction between 'natural' and 'chemical'. The water industry has been using chemicals in its treatment processes for a long time, and it adds chlorine to drinking water to prevent bacterial contamination. It also adds fluoride in the majority of the country, in order to help prevent dental health problems in the population. The use of chemicals has caused some controversy in the past and so the industry works hard to maintain an image of itself as being in the service of the public's health and as working with 'natural' technologies. The impacts of this political context were felt on the initial industry day, early on in our collaboration. As certain ideas were proposed the industry managers would run with them, 'going wild' as it were, but only up to a certain point, at which point they turned to more serious talk. If academics pushed too far in a direction that jangled these political sensitivities then the industrialists would push back, citing consumer, CCW and regulatory concerns. Academics thus had to keep pushing for exciting ideas and getting people enthusiastic whilst maintaining the trust and understanding they had built up with the water company managers by acknowledging their understanding of these issues around risk and cost.

The result of all the discussions on this first industry day was the production of several large posters covered in colourful sticky notes with ideas from the mundane and long-standing to the outright absurd. Alongside the possible applications of synthetic biology, barriers to the uptake of these technologies were also generated. It was agreed that four hurdles needed to be addressed, none of which was specific to synthetic biology. Instead, they all applied directly to long-documented barriers within the practices of the water industry and the regulatory system. The four hurdles were that synthetic biology solutions would have to be acceptable to the water industry and other regulators, technically reliable, acceptable to customers and commercially viable. In this regard, synthetic biology's promises did not do much to dent the barriers that were foreseen for innovation in the water sector, primarily because these barriers had very little to do with the academics' practices. Nonetheless, there was a lot to be hopeful about as overall the water company managers seemed largely enthused and had certainly begun using the terms of synthetic biology to talk about microorganisms and genes.

As the day drew to a close, the goal-oriented character of synthetic biology, aiming at particular industrial targets, came to the fore and the academic leaders needed to pin down what to do next for the project. So, we voted. The messy reality of social processes, the meeting and arguing that had been central to these sometimes heavy and at other times giddy interactions were now to be 'cleaned up'. The field of possibilities was narrowed radically. Selections were ultimately in keeping with the rough sketch laid out in the original proposal, which highlighted two preferred application options: pathogen detection using a biosensor, and pipe smoothing using engineered biofilms.

Thus, the day was a curious mix of 'opening up' and 'closing down'. Parts of the day involved wide-open discussion, with everyone having a say, putting forward ideas without too much worry about how they would be received. There was a profusion of neon sticky notes and the excited hanging of bullet-pointed flip charts on the walls. Ultimately, however, the project stuck with both what the academics thought would be doable within a given timeframe and was closely aligned with how they had won funding from the research councils.

> So the projects they've thought up are a compromise between what is practicable to do within academia and what they believe to be of benefit to the water industry. But you were at the workshop, there were some other opportunities that came up that one might argue might have been better projects to go for. But that's the way quite often things happen with academia. (Industry Broker 1)

This closing down of options was crucially to do with how immediate pressures and modes of valuing time spent and work done were organised through academic practices and not those of the industry.

Syncing practices

Undeniably, the RCUK funding shaped how our engineering colleagues approached their experiments in working with the industry actors to generate ideas for the industrialisation of synthetic biology. Although the project did not have to work on the specific proofs of principle proposed in the original grant application, these ideas had been developed by the academics to fit with their existing expertise, interests, industry challenges and, to them, represented 'doable problems' within the grant's timeframe. As such, keeping things open to industrial input was difficult to maintain once the reality of everyday academic life returned. The industry days had begun the work of getting synthetic biology to fit in the water industry. The academics had done a good job of synchronising the narratives and affects. But the longer-term goal of getting synthetic biology to work for engineering in the context of water was not going to be accomplished just by extending the timeframe of expectations and making everyone excited. This was because the academics served other masters and had to satisfy other more immediate concerns.

Their own goals, life projects and careers were more subject to the structures of the university than they were to the promised future of synthetic biology. They had to write papers that would do well in journals that were respected in their fields. They had to find jobs for good postdocs. They had to give conference presentations and demonstrate to the research council that the money they had spent had produced some cross-disciplinary outputs and might lead to social and economic impact. As such, their efforts in making synthetic biology work for the water industry quickly became more academic and less directly relevant to the markets and innovation spaces in which they envisaged their outputs ultimately having some usefulness.

As the project continued and progress in the laboratory was slower than hoped, priorities had to be altered. The iGEM team was supposed to be making a biosensor to detect pathogens in water and they were doing quite well. Indeed, it looked like they might be the first team at the university to go to compete at MIT and win a medal. However, their work would not be enough to satisfy funders. As such, the work being conducted on biosensors by the project leader and some of the other engineers and molecular biologists became significantly more important and time sensitive. There were some barriers here too, however. The molecular biologist on the project had been brought in for only a short period, to supervise the iGEM team and to catalogue and characterise some of the existing parts in registries that might be of use to the academics' work in designing a biosensor. It was hoped that, once they were finished, the iGEM team's work would be able to feed into the academic's work, but this was looking less likely. The academics needed something to show for their efforts, however, and so they wrote a hefty report on biosensors for the water industry based on their findings so far. They also published some of their early findings and thinking in relevant industry journals and newsletters. They put this project on the back burner whilst the iGEM team finished up, intending to return to it once it was clear how successful the students had been in getting their parts to work. The academics then began to speed up their activities in the second project on biofilms.

However, the work to develop biofilms to smooth water pipe surfaces was also flagging because of similar problems with staff turnover and difficulties in the model pipe design. The academics had to demonstrate both that it was technically feasible to grow biofilms on pipes and that their qualities could be controlled through genetic alterations. They had to show that bacteria could be designed to create a biofilm with a smooth surface reducing the friction on water passing over it. They had spent time talking to industry actors to get them excited about this, not only through the industry days, but also through other avenues. They had been involved in the discussions that we were having with the industry about the realities of water treatment facilities, and they had lots of experience themselves in what the industry would actually need for this innovation to work. There were signs that this was moving towards a more integrated way of thinking. The parts they were thinking about and beginning to work with were evaluated, early on, on the basis of how well they might fit with their longer-term industrial goals. However, as time became short, and there were problems with keeping the postdocs who they wanted to work on the project, they quickly had to push the research farther and farther away from the realities of pipes in the water infrastructure.

For example, they now needed a model pipe on which it would be easy to grow bacteria, and that would allow them to check both rapidly and regularly a biofilm's properties in their experimental set-up. Such a pipe turned out to be nothing like a pipe that would actually be found underground. First, the lab pipe would not have alternating pressures, water flushing through it at various speeds, bacteria-killing chlorine or any number of other impurities. Second, the bacteria they wanted to manipulate to grow in the biofilm would, ultimately, have to live in these conditions. However, the standard parts from repositories, any novel parts that they could design or borrow and the model bacteria themselves would not necessarily be appropriate for use in those conditions as they currently stood. In this regard, the existing standardised parts were a constraint. But the team did not have the time to engineer a new strain or 'chassis' for these conditions. So they opted for a chassis that would prove the principle and grow on the pipe, but not really represent how a final product would have to manifest within an industrial situation. They knew how these bacteria worked and they had skills in the team that would enable them to get the model pipe up and running more rapidly, but the realities of the practices of everyday academic life and its reward structures pulled the model pipe and the bacteria away from the realities of the water infrastructure. Certainly, it was a proof of principle, but it was one without much bearing on how water pipes really work. It was just too difficult – with a short timescale, too little money, changing staff and high expectations – to synchronise the industrial and academic practices in any meaningful way at this stage. Integrating industry concerns and the realities of industrial services into the material designs of their microorganisms would just have to wait until they had more time and money. Most immediately, it was academic constraints that prevailed.

Conclusion

There are promises galore about synthetic biology, and many of them are economic. It is clear that such promises are actively created to try to bring about a new kind of industrial future. Speculations about the economic power of synthetic biology are extraordinary; for example it has been argued in governance of the field that 'the value of the global synthetic biology market will grow from $1.6bn in 2011 to $10.8bn by 2016' (SBRCG, 2012: 4). The field was understood by our colleagues to be an interdisciplinary science, and offered a 'sticky terminology' through which to bring together a range of actors, whilst also promising to help them to create commercially relevant microorganisms, and so to stand a better chance of helping the engineers cross the 'valley of death' and scale the boundaries between their labs and the water industry. They wanted to be a part of bringing about new markets for synthetic biology.

Such promises and speculations are premised on a shift in academic practices. The adoption of engineering epistemology, it is assumed, will better synchronise with industrial practices for innovation leading to the better translation of inventions into profit. Synthetic biology's emphasis on changing how academics work is thus premised on the assumption that industries will be receptive to innovations coming from the field. However, there is not yet much evidence that companies are willing to shift their practices to align better with those of academia. For example, there is no discussion of how immediate industrial reward structures and long-term funding mechanisms might have to change to fit immediate academic rewards structures and long-term career advancement mechanisms better. Moreover, there is little consideration of whether companies will have to accept lower profits in order to bring about the more sustainable futures promised by synthetic biologists and demanded by governments. In both cases, the onus is very much on academics to bring about changes in the academia-industry interface to overcome the valley of death without any implications for the reconfiguration of markets or innovation mechanisms on the other side. However, academics' abilities to change the practices required to enable a 'second industrial revolution' are severely constrained by what engineering epistemology can actually do about these broader constraints on this interface and on the successful uptake of novel products. In this regard, synthetic biology has only, so far, brought the science side of the interface into serious question, and not the industrial one. Putting themselves in the service of industry might not be enough to bring about the kinds of relations and innovations towards which they are working.

In the pursuit of an understanding of barriers in this chapter we have explored a multitude of bounded things – pipes, disciplines, companies – that effected physical, symbolic and imagined barriers. The project we took part in was specifically designed to address certain imaginations of 'innovation barriers' and gaps between academia and industry. The descriptions we have produced here hint at the assemblage of hurdles, solutions, technologies and people that the project sought to bring into contact in a novel way, and through use of synthetic biology as the apparatus, to address such barriers. The desire to engineer solutions to water challenges tied together the interests of the academic and industry actors. Our colleagues worked hard to synchronise the affects and promises of synthetic biology with those in the water industry in order to recruit water company actors into the imagined future of the field. They were successful in doing so.

However, they were only able to go so far. Limitations on the industry side were difficult to bring into question, let alone transform. Everyday academic life also constrained the kinds of changes that our engineering colleagues could make to their practices. Certainly they did change how they approached the manipulation of bacteria. The engineers got the molecular biologists thinking and talking in a slightly different way. They began to mix in more standardised parts and so forth, although even here such standards did not necessarily make for an easy translation into the messy reality of water infrastructure. Quite the reverse. Moreover, the broader structures of academic life were not changed by these largely epistemic shifts. In this regard, boundaries and barriers are produced through practices at multiple levels, and must be understood as dynamic and contingent on how they fit into the routines and structures of everyday academic life. They are not fixed entities but are made and remade in scientific practices, even if they are changing.

Overall, we find that when practices are changed to reduce or remove barriers, they might simply reproduce them, excavate existing but unseen barriers and create new barriers in novel places and for potentially unexpected reasons. In this way, reconfiguring practices may serve to reorder barriers, not necessarily to reduce or remove them. Without changes in underlying structures of practices and modes of organisation of everyday academic life and industrial markets, the possibilities for enabling the kinds of changes that were attempted in our project will probably also have limited success elsewhere. Participants' imaginations of barriers were robust and resilient to change because they suited particular ways of thinking shared with the industry and fitted their hopes for the future, in which the industry would take up their innovations if only the pesky regulators and consumers would get out of the way. Yet, these perceived barriers are not necessarily the ones that needed addressing. Their focus on how regulatory structures shaped industrial investment in R&D did not allow for the fact that water companies might not leap towards long-term investments in high technologies because of how their profit mechanisms actually work. It is apparent that entrenched views on what constitutes a problem thus constrain the potential ways in which the integration of synthetic biology into industrial markets can be imagined and so brought about. Frow's (2013: 445) argument regarding the practices of valuation in synthetic biology at different levels of abstraction rings true in our situation. Such valuations 'operate on different timescales and are concerned with different objectives [...they also] shape and become embedded in the material and social worlds of an emerging field.'

It is thus essential to acknowledge that the field of synthetic biology is being introduced into complex, pre-existing sets of relations, practices and sociotechnical imaginations. The field is not being introduced into a vacuum. What synthetic biology becomes, how it is practised, how it makes sense to actors and is integrated into future visions depends crucially upon how the practices through which it is enacted are organised. Currently, the kind of synthetic biology being enacted in the UK is one shaped by a multi-scalar galaxy of practices that include the techno-politics of UK research governance, industrial markets and investment regimes, what can be imagined within contemporary discourses of global challenges, the multi-layering of context-specific regulation in the UK, EU and internationally, and so on. We have to understand synthetic biology within these complex, and historically situated 'constellations of practices' (Wenger, 1999: 126).

In the following chapter, we extend this analysis by exploring how everyday academic life was interwoven with attempts to change practices around the use and production of microorganisms. Taking a slice through barriers has helped us to see how our colleagues struggled to synchronise practices. Taking our next slice through bacteria opens up questions of materiality, agency and the co-production of academic careers, sociotechnical fields and their objects.

Coda 2
Brokering Relations

Is there a Google Translate from sociology to engineering? (Academic Molecular Biologist 1)

Our work in 'trying to collaborate' involved taking lots of different roles, in part because actors in the project had different ideas about what kinds of things we would or should be doing. That different people had different imaginations of sociological participation in science should come as no surprise to readers. But how did this play out? As ethnographers, we have to engage with issues of positionality; with relations between field-workers and their sites; with matters of identity, discipline, power, and so on. And these perhaps become more acute when read through the lens of attempts at collaboration with academic elites in positions of greater power. As we explored in Coda 1, the notion of collaboration has become entangled with post-ELSI attempts to reconfigure the relations between the natural and social sciences. As such, it is crucial that we attend to how this is accomplished and how these attempts sometimes fail, or succeed, within the specifics of local situations.

As we each acclimatised and settled into the project we fast found ourselves resisting certain attributed roles, while perhaps still performing these, sometimes unconsciously through habit, and at other times just by taking the path of least resistance. One such role was that of 'representative of the public' as we, and others, have described elsewhere (Balmer *et al.*, 2015). We also consciously adopted more desired roles, such as being a co-producer of knowledge, while ultimately failing to make these 'stick' in a substantive or longer-term fashion. We looked at how our roles shifted through the kinds of relationships we established with different actors in the project and how other actors' efforts helped to position us. For now, we want to focus on a particular role, that of *broker*. This is one specifically designed to reduce barriers and one that natural scientists and engineers, in synthetic biology at least, often attribute to social scientists. But we were not the only ones who became brokers in the project.

In an explicit attempt to manage how academia and industry were brought together, and to help make the most of these connections, an industry broker was built into the grant. The person appointed had 30 years of experience working in various roles within the water industry and, since his retirement, had been acting as a knowledge transfer consultant. This was very much a professional brokering role (Meyer, 2010) and was understood by the actor himself, and others in the project, as being a particularly difficult task. It was a big job: ensuring industry involvement in all stages of the project; advising on the industrial relevance of academic ideas; helping to shape the presentation of the project and its findings to the industry and promoting engagement from regulators. Interestingly, the role was deemed essential by the project leaders despite the fact that the academics involved had long-term relationships with industry, particularly with R&D managers, and existing knowledge of their needs. This was because they anticipated a certain unease in the industry with biotechnology:

[We] have a good feeling, having worked with industry for a number of years, about what they're comfortable with and what they're not comfortable with, and with these types of [synthetic biology] ideas they would definitely not be comfortable. (Academic Water Engineer 2)

Brokering was therefore organised around trying to ease the industry into a more comfortable relationship with synthetic biology. This is how we, too, were assigned the role of broker alongside the industry professional. Our expertise as social scientists was drawn on in order to aid in the management of the project and for shaping the conduct of the relations between the engineering worlds of academia and industry. Our positioning as brokers comprised trying to work out how to challenge the conservatism of industry so as to open up water engineering to synthetic biology practices. As we have described, the industry was understood to be unadventurous and risk-averse because the water regulators were seen to be too heavy-handed and the public ignorant of the true costs of water provision and treatment. This was framed as a 'social problem' and thus under the purview of our expertise. As one colleague described:

I don't think we can deliver on synthetic biology in the water industry unless we look at the social aspects. ....it's not science in isolation, it can't be....the social aspect is forming a really integral part of teaching me how to think beyond the lab. I think you can provide us with a skill set of how to engage with industry, on this idea of barriers to implementation and things like that. (Academic Environmental Engineer 1)

Our role then was seen to be tangled up with ensuring that this was not 'science in isolation' and to help academic scientists and engineers to 'think beyond the lab', blurring the lines between broker and teacher.

To understand how we might do this, the project leader talked to us during meetings and so on, asking questions to try to find out what we had been doing. Of course, we also volunteered presentations, wrote reports and actively engaged colleagues with our research. However, our collaborative partners also drew on conversations outside of working environments, in 'coffee shops or whatever' (Academic Environmental Engineer 1), where more informal talk could take place and mutual understandings could be built. In these contexts, we could find similarities in our experiences of being academics, talk about the challenges of working at the university, think about our careers, and so on. Things could be said here that were more difficult to say in a project meeting.

Our critical engagement with the project thus became entangled with how we were building trusting relationships in a more everyday fashion. Through these various avenues the project leader, and some others in the team, began to see value in thinking more reflexively about the organisation of the work and took on board suggestions we made as regards interactions with industry, recruiting and managing the iGEM team and thinking about 'social problems' like the 'ignorant public'. This was also mixed up with longer-standing ways of thinking, habits and structured practices. As such, our brokering role navigated relations not only between academia and industry but also engineering and sociology. In this regard, it is important to remember that how social science is understood in scientific practices is not just about how we enter the laboratory, but of how our professional academic and, to some degree, personal lives are woven together through efforts at collaboration that mix progress, resistance and recalcitrance. That everyday practices like getting coffee can be crucial to opening science up more informally warrants further exploration, in STS at least, so that we too can 'think beyond the lab' when it comes to collaboration. Everyday life in experiments with post-ELSI entanglements has to figure in how we develop such epistemic, interdisciplinary practices.

At the same time that the project leader sought not to do science in isolation, she also struggled to understand what exactly this might mean. As she described it, 'I don't know how to go beyond building a biosensor.' In being given responsibility for the 'social aspects', we were imbued with an expertise that the engineer in question felt she didn't have. We were seen as being able to provide a 'skill set' of how to 'engage with industry' and as being able to help scale the innovation barrier. So whilst there were opportunities in which novel conversations could be had, particularly in the informal everyday settings of collaborative projects, there were also forces constraining what kinds of things could be done 'beyond the lab'. The conceptualisation of barriers to innovation and the way in which our expertise was understood meant that we were still expected to take on a kind of professional brokering role. We pushed at the limits of this, trying to examine what kinds of reflexivity might be brought to bear on the concept of 'barriers' to innovation, but it was a struggle. It became more difficult to resist through the way in which we were positioned between a kind of pastoral and PR role in 'giving bad news' and presenting risks in the right way to industry actors:

> Andy Balmer: if something is risky in this project or in synthetic biology more generally, what should you do about it?

> Academic Environmental Engineer 1: I was hoping that's where the social scientists could come into it, to give us the tools to deal with that.

> AB: What do you mean by tools?

> AEE 1: Well if you come up with something that's really risky, how do you present that to someone, in the industry for example? There are ways you should present it and ways you shouldn't present it. So I guess some information on how to present that. It's almost like giving bad news. I mean, that's something I don't know how to do.

It was not clear what sociological practices would enable us to meet these expectations. Perhaps being a social scientist implies to engineers, such as our colleague, that we possess unique *social* skills for use in social situations or that we can learn to speak the language of industry in a way that they cannot. More pointedly, our place in the project was seen as both an experiment that might just fail, ending up in sociological outputs that were irrelevant to the engineers. Further, what would constitute success was sometimes just conceptualised as our work being the means to a pre-determined end, namely that we would help to clarify how our colleagues should act and speak in order to allay the industry's fears of synthetic biology:

> We're very intrigued to see what comes out of [the sociological partic-ipation]. With my pessimistic hat on, we'll do lots of these interviews and be watched a lot and then something really woolly and vague will come out at the end. I think a nice bullet point list of do this, don't do that, do this the next time, make sure you understand this, set these boundaries, get this common understanding, whatever it might be coming out the end. It could be something really tangible that says how we actually make [the project successful]. (Academic Water Engineer 1)

The issue of sociological language also came up in our engagement with the industry broker. For him, the problem we had faced when inter-acting with industry actors was that we did not talk appropriately:

> I think one of the issues might be the language you use when we engage with them. It needs to be, certainly less academic, less esoteric. I'm not saying that the social aspects are esoteric, but from their perspective it's not part of day-to-day business. (Industry Broker 1)

Although assigned the role of official project go-between, and holding responsibility for facilitating industry engagement, troubles with the interface were located as beyond the industry broker's remit and were left unsettled. His emphasis was on the need for academics, and perhaps in particular we social scientists, to change our practices to better align with industrial needs. He wanted, for example, us and the engineers to change how we wrote grant applications, conducted work and talked about research. The broker rarely discussed how the industry should change, except to say that it could be more adventurous in its research agenda. Even then, he would always blame this lack of adventure on the public and regulators for hamstringing industry with unnecessary bureaucracy and fixed pricing. As a result, how exactly he hoped to bring about change was opaque.

This differed from the academic perspective. Our academic colleagues argued that the project *was* informed by industry needs, from having drawn on their own experiences of previous research relationships, and that the project had been designed to be collaborative from the start. To emphasise this they pointed out how the project's timeline was anchored in the three 'industry days' that were funded, and would pull in actors from across the water companies to discuss the project and its aims. They also saw 'the social' as being central to the day-to-day business of industry, and as being the root of innovation failure. For them, the 'social aspects' were far from being an esoteric concern.

Our position as brokers in the project perhaps became more prominent because of this. The economic and regulatory barriers were understood, partly because of how the industry broker framed the issue, as being insurmountable on the industrial side of things. So if the barrier was going to be scaled it would be on the academic side. Further, as researchers understood to be experts in 'social aspects', the onus was distinctly on us to help to do this. Indeed, our sociological methods themselves, which regularly took us into industrial sites and brought us into dialogue with industrial actors, helped to shape how we were understood in this regard. We would bring back stories from the field, present interview data and ask questions of the academics based on what we were observing in the industry. This all meant that in a very practical, everyday way, we *were* brokers, ferrying information back and forth. However, it meant that this important work remained largely invisible until we published papers that could be counted as academic outputs.

As we developed and presented our work in conference presentations and papers it obviously took on a more distinctly sociological flavour. We coined new terms, adopted existing ones, framed the data differently, organised the work around different questions and so forth. One of the first papers we published was discussed in a meeting about another topic. A colleague from the project said that they had read the paper and did not understand it. In the midst of this conversation, Andy tried to explain what the argument was, but there was a sense that this had not clarified things terribly well. In part, the problem was that our diagnosis of 'barriers' that we developed in the paper (Molyneux-Hodgson and Balmer, 2014), could not easily translate back into the research situation from which it had emerged. This was, perhaps, because Andy was not adequately equipped, intellectually or practically, to translate the academic paper into something understandable in a few minutes of talk without preparation. Or, as another engineering colleague had requested, as a few bullet points of dos and don'ts.

The paper we wrote tried to understand the ways in which academics approached industrial relationships, and argued that this might actually be partially constitutive of the barriers themselves, by virtue of how a range of interconnections of regulatory, governance and innovation practices shaped their actions. So perhaps the argument was esoteric, or perhaps just wrong. Or maybe there was another barrier here, this time one of language or conceptualisation.

Whatever the reason for their difficulties in understanding, there was also an important issue regarding power. The expectation was clearly that our academic language and practices should serve the role of brokering relationships with the industry. Sociology was expected to provide something akin to a pidgin language, and to serve the role of brokering in a very literal sense of translation. Moreover, it was implicit that the engineers and scientists in the project should be able easily to understand what we had to say. The common use of terms like 'esoteric' or 'complicated' to designate the jargon in social science signals a certain expectation that scientists and engineers do not have time to learn the language or practices of social science and that it is rather we who should make the effort of translation. There is therefore a question, and one that is practically resolved in each situation, regarding power as concerns the time and effort differentially invested in making possible and maintaining more substantive collaborative relations and dialogue. What we could do with the brokering role in this situation was constrained by how this particular issue regarding power manifested.

There is also an ontological implication to scientists and engineers' expectations of easy access to social science: that social life itself should be transparent, and that our language should thus mirror that of everyday life. Of course, to some degree, and dependent upon your theoretical disposition within academia, this is true. However, sociology, STS, anthropology and so on, have developed a range of concepts to describe and relate things for which there are no existing words. This draws on the long history of the social sciences, but also on philosophy, history and even the natural sciences themselves. Some of these terms filter back into everyday life, so that people talk about the unconscious, social class and so on without needing an academic training. Ultimately, however, this ontological issue relates to relations of power in that it is often implied that social science terms are 'made up' and that natural science terms are reflections of material reality. What there is and how this affects how we relate to each other are inevitably entangled.

Ultimately, we found being brokers on the project, alongside the professional industry broker, to be a difficult role. The entanglement of power relations, language and practices within our efforts to construct collaborative relationships were indicative, to some degree, of a larger problem. The position of the social sciences in such interdisciplinary encounters, in universities and in governance of science and the economy, is subject to a range of constraints that are largely outside of our immediate ability to change. Our relationships in personal and everyday academic life did act as a space where there was more flexibility, but to a limited extent. We have to do more in STS to examine how power relations manifest in experiments with collaboration and to create new kinds of equipment to challenge them.

3
Bacteria

> Most people don't realise it, because they're invisible, but
> microbes make up about a half of the Earth's biomass, whereas
> all animals only make up about one one-thousandth of all the
> biomass. (Venter, 2005)

Introduction

Bacteria are everywhere. We heard this regularly throughout the project.
For the microbiologists and molecular biologists we met, bacteria were
absolutely central to their daily research activities. These scientists were
immediately responsible for the laboratory work and for engaging practically with living bacteria. On the face of it, they were the people most
involved in trying to realise the academic team's hopes of demonstrating
a proof of principle that synthetic biology techniques could be used to
solve water engineering problems.

But bacteria were also central to the work of a number of our engineering colleagues in other disciplines. For the water engineers bacteria
were a feature of everyday work because they are present both in
clean water distribution systems and sewerage treatment systems. For
the chemical and environmental engineers, bacteria had also already
become part of their research in one way or another. They all came to
the project team having become knowledgeable about bacteria, some
having developed quite significant expertise about the strains with
which they worked, their manipulation and in scaling up their production for industrial use.

Bacteria, then, were not unknown to anyone working on the project.
Indeed, in our own sociological research, bacteria were fast becoming a
part of our everyday lives too. In accordance with the practice in some

STS methods to follow the actors, we at times switched from following our academic colleagues to following the microbes with which they worked. In doing so, it fast became clear that bacteria were central to trying to enact the field of synthetic biology in the water industry and in the university. Of course, they are also at the centre of bringing about the field more broadly. As we have already described, proponents of synthetic biology intend to standardise the practices for engineering bacteria, so that they are more predictable and controllable and their design is no longer bespoke. This is how they mark a distinction between their own approach to manipulating bacteria and those used in other fields. In our project we regularly encountered such distinctions being made by actors of various sorts.

Whilst each person in the project was entangled with bacteria in one way or another, and whilst they all worked with existing standards of various kinds, they also each had their own distinct set of practices, ways of thinking and hopes for the future. All of which shaped how they worked with bacteria. Further, although the project had 'synthetic biology' in the title, and we were explicitly concerned with making the field relevant to the water industry, none of our actors adopted the emerging norms of synthetic biology wholesale. They were all, to various degrees and in a number of ways, engaged in making their existing practices and those of synthetic biology fit together. How they talked about bacteria and their engineering practices were crucial features of this process.

The central question that we address in this chapter regards the ways in which the ontology of bacteria in synthetic biology was stitched together with pre-existing concepts, practices and problems in the work of those researchers who had been brought together by the funded project. The promises and practices of synthetic biology that our colleagues encountered at the core of the field did not simply supplant existing practices here in the periphery. Plenty of things were borrowed from the core. However, instead of replacing local practices, they were negotiated, transformed and reinterpreted alongside and through the extant elements of everyday work in our local situation. In this regard, our actors had to work to make bacteria sensible through the practices and promises of synthetic biology given both local constraints and opportunities to change their practices and those of the water industry. To begin to put together this argument we first describe the concepts of 'bio-objects' and 'bio-objectification' in order to apply them to the research being conducted at the core of the field. This helps us to understand how it is that core actors are generating novel ontologies of bacteria. We then move back into our everyday work at the periphery, and begin to explore how it was that these core ontologies fared within local practices.

Bio-objects and bio-objectification

A number of STS scholars have studied the co-production of biological objects, sociotechnical systems, everyday practices and governance regimes. Such 'bio-objects' (Vermeulen *et al.*, 2012) might be anything from the first synthetic cell, constructed by scientists at the J. Craig Venter Institute (JCVI), to an IVF embryo frozen in a tissue repository. Whatever they are, scholars working in this area argue that bio-objects are formed from the 'generative relations' between technical, economic, political and social practices (Metzler and Webster, 2011). The processes of creating bio-objects through such generative relations, it is argued, are best understood at the level of everyday technical practices, which work to 'objectify' life so as to tame and control it, and so to render it amenable to various other purposes. This work is conceptualised as 'bio-objectification' (Holmberg *et al.*, 2011). It is the process through which life-forms are made into objects and attributed with specific qualities. Work on bio-objects has a potent resonance with the emerging concern in STS regarding the enactment of ontologies within practices, as we briefly described in Chapter 1. What bio-objects are has to do with how they emerge from the practices through which such productive relations are made and negotiated.

For example, this can be seen in the related area of systems biology, which deals largely with computational models of microorganisms, and which plays a major role in the development and current practices of synthetic biology across a range of situations. Vermeulen (2012) argues that the technical practices of systems biologists modelling cellular processes in software programs generated a new bio-object, which became known as 'the silicon cell'. She shows that the decoding and recoding of cellular life into silicon is a process of objectification of the silicon cell, one that endows it with particular attributes making up a particular kind of identity. The ontological status of the silicon cell is thus enacted from within systems biology's computational practices.

Moreover, in the case of systems biology, these processes of bio-objectification not only reconfigure the boundaries of life but also transform the organisational boundaries of scientific disciplines. Vermeulen describes the history of entanglement between the bio-objectification of the silicon cell and the concretisation of systems biology as a distinct field of study and – eventually – as an object of governance itself. For example, the objectification of the silicon cell was later influential in establishing various European funding and governance initiatives, institutions and academia-industry consortia, all of which were geared towards the consolidation and international integration of systems biology practices, governance and innovation. Vermeulen articulates how different phases in the objectification of the silicon cell bio-object were related to different organisational phases in the transitions from wet biology, through systems biology to 'big biology' as information and communication technologies were further integrated into the biological sciences. Bio-objectification thus has implications for science governance.

In that sense, the processes of bio-objectification are also co-productive. Bio-objects are created from out of generative relations, and at the same time lead to changes in those relations, potentially creating new sets of institutional, political and individual entanglements. This also means that previously stable definitions of certain terms and the performance of certain practices may become more or less contested and in need of further governance. Describing these kinds of messy entanglements, Holmberg *et al.* (2011: 740) propose that:

> As a consequence of these novel relations, the boundaries between human and animal, organic and nonorganic, living and the suspension of living (and the meaning of death itself), are often questioned and destabilized, and their identities have to be negotiated and (temporarily) stabilized, and so given an identity. What is common to what we call bio-objects, is that they all in various ways challenge conventional cultural, scientific, and institutional orderings and classifications.

Synthetic biology is an excellent case in this regard, and the first synthetic cell created at the JCVI is a good example of technical practices of bio-objectification, the co-production of new generative relations and the disruption of existing classifications. As the JCVI team describe on their website:

> Genomic science has greatly enhanced our understanding of the biological world. It is enabling researchers to 'read' the genetic code of organisms from all branches of life by sequencing the four letters that make up DNA. Sequencing genomes has now become routine, giving rise to thousands of genomes in the public databases. In essence, scientists are digitizing biology by converting the A, C, T, and G's

of the chemical makeup of DNA into 1's and 0's in a computer. But can one reverse the process and start with 1's and 0's in a computer to define the characteristics of a living cell? We set out to answer this question. (JCVI, 2010)

Their answer was to design *in silico* a modified genome based on the sequence of the bacterium *Mycoplasma mycoides,* to synthesise the genome in yeast cells and then transplant the whole genome sequence into 'recipient cells' of *Mycoplasma capricolum* (Gibson *et al.,* 2010). As they say, they then 'booted up' the genome and the self-replicating *Mycoplasma mycoides* JCVI-syn1.0 cells were created. Reporting on their research in the *Wall Street Journal,* Craig Venter and Daniel Gibson described various relations between the concepts of 'natural', 'synthetic' and 'artificial':

we [did not] create life from scratch. We transformed existing life into new life. We also did not design and build a new chromosome from nothing. Rather, using only digitized information, we synthesized a modified version of the naturally occurring *Mycoplasma mycoides* genome. The result is not an 'artificial' life form. It is a very real, self-replicating cell that most microbiologists would be unable to readily distinguish from the naturally occurring counterpart without the aid of DNA sequencing. (Venter and Gibson, 2010)

In this account of their work, the authors explicitly address the question of how their research might or might not have implications for the status of the bacterium as natural, synthetic, real or artificial. In drawing boundaries between these terms and in attributing some rather than other of these qualities to the bacterium, Venter and Gibson objectify the organism in particular ways. They enact its ontological status through these practices of objectification. Whilst doing so, they implicitly demonstrate how instabilities in such definitions and broader sets of relations that result from the creation of the organism must be settled through such acts of description, and are not simply given in advance of their bio-objectification. They go on to quote President Obama's response to the announcement of the synthetic microorganism, in which he refers to the work as 'a milestone in the emerging field of cellular and genetic research known as synthetic biology.' In this way, much as with the case of systems biology's silicon cell, the bio-objectification of the *M. mycoides* JCVI-syn1.0 cells was importantly coupled to the consolidation of synthetic biology as a distinct scientific field.

Subsequently, a range of other actors engaged in discussions about the implications of this work for classifications of life and for the governance of these sociotechnical practices of genetic engineering. The Action Group on Erosion, Technology and Concentration (ETC), an international campaign group concerned with socioeconomic and ecological issues surrounding new technologies, dubbed the new organism 'Synthia' (ETC, 2009). The group questioned whether Synthia represented a panacea or Pandora's box (ETC, 2010), citing the profit-driven nature of the field, the potential for synthetic organisms to be used as biological weapons and argued that Synthia represented a threat to natural biodiversity. They also drew attention to the connections between the JCVI and multi-national energy companies such as BP and Exxon. The JCVI group themselves insist that they regularly engage in discussions about the 'ethical and societal implications of their work,' stating that 'continued and intensive review and dialogue with all areas of society, from Congress to bioethicists to laypeople, is necessary for this field to prosper' (JCVI, 2010).

As such, the announcement of the cells and their identification as Synthia, along with the subsequent debate and ongoing efforts to 'engage' with the public, evidence the potential for bio-objects to unsettle the very categories and generative relations through which they were constituted. They do so by opening them up to contestation and possible reconfiguration. This is well understood by scientists and engineers working in the field of synthetic biology, though generally not in these terms. They frequently draw attention to the 'implications' of the bio-objects that they are creating, though less frequently to their practices and relations with industry. In these ways actors that might describe themselves as synthetic biologists routinely invoke the future as part of their own narratives of bio-objectification. They also draw attention to the field itself as having 'ethical' and 'social implications.' However, they largely refer to synthetic biology as a metonym for the host of objects and products that they envisage developing.

Importantly, the ways in which synthetic biologists describe such issues of governance in relation to their bio-objects is not disinterested. Rather, it is often explicitly connected to the perceived need to ensure that synthetic biology does not lose momentum because of public contestation, fears or controversies, and thus to the requirement to make sure that synthetic biology practices create economically successful bio-objects. What is generally less explicit in their talk is the relation between their accounts of bio-objects, their accounts of synthetic biology as a field in its own right and their own practices of self-identification *as* engineers, molecular geneticists, synthetic biologists or whatever. In this regard, they are less likely to invoke themselves as being at stake, of how their own interests play a role in the configuration of the field and future narratives and of the ways in which their careers progress, or not, through experiments with synthetic biology.

In the following two sections, we begin to unpick this relation by evidencing the entanglement of bio-objects, fields and disciplinary identification, and situate these issues more mundanely within everyday life and academic practices. First, we examine how our colleagues in the project described the initial shift in their conceptualisation of bacteria. Before the advent of synthetic biology, our colleagues who had trained as water, environmental or chemical engineers had already intentionally worked to shift the ontology of bacteria as part of their research into engineering problems. We conceptualise this first shift as 'bacterial emergence', in which bacteria became more active participants in the construction and possible solution of engineering problems. As part of this process of bacterial emergence, we begin to see how our colleagues' disciplinary identities were also in a process of transformation as they sought to shift their practices in regard to microorganisms. To further evidence the ways in which bacteria figured as bio-objects in the pre-existing practices of the project team, well before synthetic biology appeared on the scene, we describe how bacterial emergence occurred within some of the sociopolitical problematisations of innovation, governance and academic research that we documented in Chapter 2. This helps to demonstrate how bio-objects such as bacteria are invested with biovalue as part of sociopolitical discourse in governance, industrial and academic practices. It also leads us into a more detailed consideration of how our engineering colleagues then sought to fit synthetic biology into this mix of existing technical and political practices, through their everyday work of research and professional development. In this way, we track the co-production of bio-objects into everyday life.

Narratives of bacterial emergence

During interviews with our colleagues, in observations of everyday work conducted in the project and at conferences we attended, it became increasingly clear that academics with a background in chemical, water or environmental engineering tended to share a certain vision of the world as a series of connected processes, each of which could be modelled – given certain constraints and acknowledged assumptions – in order that it could be optimised. Such a perspective was clearly invoked when they discussed their past work and education, and was similarly present when they outlined physical or chemical engineering problems involving some kind of biological phenomenon. For example, one environmental engineer described an event from early in her career. As a young scholar, our now senior engineering colleague had modelled a water engineering system that involved a microbiological component. In that model, she had treated the bacteria as inert particles. The explanation she gave for why she worked with bacteria in this manner was that it had simply worked well enough for the model to be reasonably accurate given what was required of it, or as she put it:

> We tried to ignore the biology as much as possible, because we were, well I was applying inert particle technology techniques to a biological system, which was the first time that they'd ever tried to measure these changes in particle size of bacterial floc and things. And for the modelling system it was easy to assume that they were inert particles rather than biological things that grow and divide. So, reasoning, we said the time is too short for them to do anything (we were measuring them over maybe two hours or something), which now I know is probably not true. But for the purposes of that [experiment] we said, 'Well let's just look at them as being inert, apply the techniques...' and we got some really sensible results. (Academic Environmental Engineer 1)

This excerpt is characteristic of the ways in which many of our engineering colleagues articulated a first shift in their practices of working with bacteria, resulting in a change in the ontology of bacteria, which we term 'bacterial emergence'. In the above quotation, bacteria are first described as 'floc', a technical term from water engineering that refers to small clusters of filamentous bacteria, produced by the addition of coagulating agents during water treatment processes. In the second instance, this floc is treated as an 'inert particle' in order to model the system using particle technology techniques. Finally, the bacteria become – as the engineer speaks from her perspective now – living things that 'grow and divide', and so change over time. One significant point to draw from this is that the early ontology of bacteria as being inert was entangled with a specific problem, an engineering problem, which provided a set of practices in which the bacteria were enacted as particles rather than, for example, colonies or cultures, as commonly happens in molecular biology. In this specific case, the ontological status of the bacteria as inert and particulate made sense within the epistemology of the computational modelling of a dynamic water system and helped to produce 'really sensible results.'

Such early-career engagements in which the biotic was treated as particulate or inert were quite common amongst our engineering colleagues. However, as with the example above, their current approach to bacteria was markedly different. This was true whether they had since grown closer to molecular biology or synthetic biology as a set of practices and terms through which to work with bacteria. To whichever disposition they had since become attuned, the similarity was in how they now conceived of the biological realm as being more active. Though many of our engineering colleagues shared an historical vision of organisms as mathematically predictable and inert, they also had in common a narrative of bacterial emergence, a story in which microorganisms had become more actively significant as regards some particular engineering challenge, which helped to bring about new ontologies.

In these stories of emergence, the bacteria were no longer inert particulate entities and instead became 'active', 'living' or vital in some way. Each of these new attributes invoked relations between bacteria, their surroundings and the problems with which engineers were now engaged. Asked why she now chose regularly to work with bacteria, one researcher trained in environmental engineering put it this way: 'They chose me!...It turned out bacteria were the main catalysts of the reaction [I was studying] so I had to learn to work with them' (Academic Environmental Engineer 4).

A chemical engineer on the project, who sometimes also identified as a synthetic biologist, described his experience in working with bioreactors as part of his postdoctoral work:

> I set about learning a lot more biology because I thought that some of the problems I was facing I needed to understand more about the biology than about the fluid mechanics of what was happening in a bioreactor. (Academic Chemical Engineer 1)

The engineer went on to describe how he had then begun a life-long process of becoming immersed in the culture of molecular biology: reading papers, tinkering in the lab, attending conferences and so on. Along this journey, the engineer described a series of transformations in his work with microorganisms, from a focus on bacterial roles within reactors, through examining bacteria as 'processes', to investigating how to 'manipulate them', and 'make products with them'. Describing this move away from the industrial concern with water treatment, he felt:

'As an engineer I'd like to make something rather than treat something. So I started learning how you made things with biology rather than how you rendered them harmless.' Ultimately, this journey led him to synthetic biology, which now offered a chance to control the active and unpredictable bacteria with which he had learnt to work.

Bacterial emergence is characterised by a shift in ontology through which bacteria become a resource for engineers, a new component and process that can be used to amend existing models and systems so as to solve an extant problem or to create novel products. The bacteria, previously inert in a model, or invisible in a system, became significant to environmental, water and chemical engineers in as much as their emergence could help them to understand and so to solve a difficulty in their research. It was specific problems of water, chemical and environmental engineering in which bacteria initially became sensible as living agents to engineers in our project.

Such stories of bacterial emergence implicitly invoke processes of bio-objectification as bacteria are conceptualised as different kinds of bio-objects, and attributed with different qualities. The ontology of bacteria changes as problems, experimental apparatus, conceptual schemes and so forth change across spaces and over time. There are differences between bacteria as floc, inert particles, living and growing organisms, processes and products, which are not simply different ways of describing the same thing found in various contexts, but are ontologically distinct things by virtue of their enactment within different practices.

In each of the contexts of water, chemical and environmental engineering research, the bio-objects are bacteria, but this is an accomplished sameness. What it means to say that they are bacteria differs according to the practices of making them known, identifying them as such and using them for other purposes. We can also see how these different practices of bio-objectification unsettle distinctions between living and non-living, organic and inorganic. For example, within Academic Environmental Engineer 1's descriptions of her earlier work as compared to her current work, bacteria were attributed with different qualities that explicitly referred to them as non-living (for example as particles in a model) or as living depending upon the role that the bacteria played in the knowledge-making practices in which the engineer was engaged. Bacteria would seem to live and die at the whim of human practices.

However, bacteria – as clearly described in these very accounts – are not simply passive receptacles willing to hold any definition or to be ascribed any qualities we might wish. Instead, they themselves actively constrain and make possible different enactments. They take different forms within the epistemological practices used to objectify them. As Academic Environmental Engineer 4 said, 'They chose me!' Our position on this ontological question is thus that the bacteria do become different kinds of things as practices shift from context to context and over time, but that they are also participants in those shifting practices. As articulated by the same engineer in saying, 'I had to learn to work with them,' bacteria can be active agents in the construction of problems, in the work that humans do to address those problems and in creating possible solutions to them.

Through describing these narratives of bacterial emergence, we have shown how our engineering colleagues had already worked to change their practices for working with bacteria long before synthetic biology began to develop. We also posit that the bacteria themselves played a role in this shift and we have begun to evidence how engineers changed their appreciation of the value of bacteria by reference to the problems on which they were working. In the following section, we continue to articulate the bio-objectification of bacteria but now locate this process within the contemporary sociopolitical context of the water engineering, innovation and markets described in Chapter 2. In this way, we start to describe how it was that synthetic biology terms and practices emerged against the backdrop of these more long-standing ontological shifts.

The sociopolitical context of bacterial emergence

Bacterial emergence was importantly enacted through engineers changing their practices to try to understand and so to solve socio-technical problems. As their narratives of changes in the ontologies of bacteria over their professional careers got closer to the contemporary research in which they were engaged, we found that both academic and industrial engineers were indeed grappling with some obstinate problems in which bacteria now prominently figured as active agents. From fixing leaks in water pipes to dealing with fats, oils and greases (FOG) in sewerage systems, such problems were difficult to deal with because they involved a range of political and material factors, some of which conflicted with each other and could not easily be reconciled. What became clear in our ethnographic participation in these everyday spaces was that, at the nexus of academia and the water industry, bacteria had begun to play a role in trying to overcome some of these conflicts.

One good case in which bacterial emergence was being negotiated was in relation to a problem with sewer sedimentation. The problem involves the way in which sediment – the collection of material on the inside of water pipes – accumulates, erodes and becomes detached from pipe surfaces. Once it has detached it enters the water stream and may end up wherever the water ends up. This was an important problem on which several of our colleagues in water engineering were working. Although it was not part of how they were developing synthetic biology tools for water problems in this project, it was closely aligned in that it concerned how bacteria were understood in water systems. Moreover, it was part of their long-standing efforts to get industry actors to take more of an interest in the microbiological features of the water infrastructure. They used laboratory experiments with model pipes and field experiments with clean and waste water street pipes to try to conceptualise, quantify and ameliorate the problem of sewer sedimentation, all within a series of regulatory and market structures.

Sewer sedimentation has recently become a more important problem for industry-based water engineers, and so also for academic water engineers, for a number of reasons, not least because of the EU Water Framework Directive (WFD, 2000). As we briefly described in the previous chapter, there are a number of layers of water regulation. At the EU level, the WFD has recently placed increased emphasis on the quality of water in surface water bodies such as rivers and lakes (Molyneux-Hodgson and Smith, 2007). This change in EU regulations has meant that water services across Europe have had to meet new quality criteria, involving, for instance, the limitation of the concentrations of particular pollutants. In high sewage flow periods (for example during storm events) sediments from the insides of sewerage pipes may become detached, flood out and mix with natural water courses, creating environmental pollution. Falling short of EU regulations on such kinds of pollution meets with heavy sanctions, and so a variety of actors have been enrolled to negotiate industrial responses to this change in regulation. On the clean water side of things, the flow of water through pipes is similarly important since increased amounts of material in water directly affect the friction placed on water, and so the amount of energy required to pump water through pipes. As sediment accumulates and detaches, particularly during periods of heavy demand, it can appear in drinking water, thereby affecting consumer satisfaction.

As we also described in Chapter 2, consumer satisfaction has become an ever more potent measure in the governance of the UK water industry, particularly since the introduction of the service incentive mechanism. One effect of the SIM is that the number of calls of complaint made has a direct impact on the financial rewards received by the water companies. Therefore, the chief concerns of the industry, as regards consumer satisfaction, are water quality (to minimise complaints, and so avoid financial penalties or earn rewards) and efficiency (to lower costs of service provision and so to keep down bills as well as to increase profits).

Water company R&D managers are charged with facilitating the reduction in costs and with helping to ensure that provision of clean water is maintained and complaints kept low. In developing technical solutions to some of the obstinate problems of the industry, R&D managers must take into account these connections between material organisation of the sewerage and water systems, profit and costs, the SIM, consumer satisfaction and a number of other technopolitical factors.

The problem of sedimentation is one example in which industry-academia collaboration is being pursued in order to try to improve water services, reduce costs and penalties and ensure that water companies maximise their receipt of financial incentives from regulatory bodies. Both academic and industry engineers use modelling to characterise the flow of water through systems in order to predict demand, improve quality, reduce complaints and prepare for high-risk events such as storm overflow. Such models have traditionally taken sediment into consideration, but have conceived of it as having the characteristics of inorganic particles and as being uniform in nature (Banasiak *et al.*, 2005). Thus, the models, until more recently, have fitted the traditional epistemology of water engineering and have treated sediment as 'inert' in the same way as described in our academic actors' accounts of bacterial ontologies prior to their experiences of bacterial emergence.

However, a number of academics and some industrialists are actively trying to shift this epistemological construction towards a conceptualisation of pipes as full of microorganisms that might actively impact on the physical characteristics of the system and the processes of sedimentation. This shift is, in part, a direct response to the changes in governance of water services that have imposed stricter controls on water quality and incentives to reduce consumer complaints, as described above. The 'inert' sediment models have become strained and no longer sufficiently accurate for the technopolitical demands placed upon them. The models are struggling to predict problems with sediment disruption and facilitate improvements in service. As such, there is a need for more accurate models, which is creating a need for more information, in particular about the biotic character of the pipes. At the nexus of academia-industry, the problem of sedimentation is beginning to be addressed by examining the ways in which bacteria contribute to the physical processes occurring in the pipe network, for example, by trying to understand how bacteria and the biofilms they create on pipe surfaces are involved in sediment accumulation, erosion and detachment.

One experimental programme, conducted by a co-investigator on the project, has been concerned to investigate the effects of varying the microbiological presence within sediments as regards water flow. This is done in the model pipes by controlling, for instance, oxygen availability, which in turn influences the density and characteristics of the bacterial community, or biofilm. The bacteria grow and interact with the sand and silt as part of the small-scale, physical model pipe experiments. These experimental apparatuses provide data in order to generate equations that can be plugged into 'virtual model pipes' for better prediction of water flow in real pipes. In this way, over the past decade or so, academic studies have begun to characterise the living, biotic world of pipes, and this has begun to alter how industry conceptualises the role of bacteria in the water infrastructure, and their future potential to solve some of the problems.

This was visible in our interviews with water company R&D managers, who very clearly connected several of the generative relations we have already identified in how they understood the significance of bacteria for their work.

Andy Balmer: Has the way that people in your company think about bacteria changed over your career?

Water Company R&D Manager 2: I think we are more aware of them.

AB: What's driven that awareness?

WCR&DM 2: I think a drive to improve regulatory compliance. I think there's an understanding there's only so much we can do to improve compliance, unless we start to deal with bacteriological issues.... I think we're more strongly aware of the biological and bacteriological processes that occur within our raw water resources, for example underground in aquifers, in terms of release of manganese from rock due to anaerobic conditions, and the part bacteria play in generating those anaerobic conditions in the removal of nitrate in aquifers. I think we're also starting to move, as an industry, to more sustainable, biologically based treatment and also in situ treatments.... So, I think our view is changing and our willingness to exploit biological processes is great.

R&D managers, such as in this excerpt, regularly linked bacterial emergence to anticipated futures in which biological phenomena like bacterial processes are exploited for capital gain and improved service provision. As such, bacteria have begun to take on 'biovalue' as part of the current and anticipated future of the 'bioeconomy' (Birch and Tyfield, 2013; Rose, 2007a; Rose and Novas, 2005; Vermeulen *et al.*, 2012; Waldby, 2002). This happens as they are bio-objectified through the generative relations between regulation, commercialisation and the academic study of the water infrastructure. We return to this point in the following sections as we examine how our project colleagues sought to capitalise on the increasing visibility of bacteria in the water industry by linking the organisms' potential biovalue to their plans to develop synthetic biology solutions to water industry problems.

As we argued earlier, we can also see within this sociopolitical case study how bacteria themselves have played a key role in their particular form of emergence within this situation. Their own biological and ecological processes, forms of organisation, actions in their local environments, and resistance to the mechanisms used to try to control them, are all features of the co-production of the problem of sewer sedimentation in the form it has taken and of how solutions to it are posed. The co-productive relations between material actors (bacteria, water, pipes, and so on), sociopolitical orders (the SIM and WFD), economic structures (privatisation, the water company monopolies, biovalue), and technical practices (academic water research, industrial R&D, measurement equipment, and so on) were entangled in the formation of the problem of sewer sedimentation, the bio-objectification of the bacteria, and their accruing biovalue.

In the following section, we turn to the connection between this bio-objectification process and practices of disciplinary identification. In fact, we can see this in the accounts of bacterial emergence that we have described already. For example, such stories and talk about particular engineering problems were part of how our colleagues distanced themselves from other engineers who had not yet awakened to the active nature of the microbiological realm. Asked why other engineers hadn't experienced this process of emergence, one colleague explained, 'I think often the biology gets ignored because it is something [other engineers] don't understand, or it's just more complicated than they think it needs to be to do what they have to do' (Academic Environmental Engineer 1). In this regard, our engineering colleagues narrated stories in which the ontology of bacteria differed by virtue of working within changing sociotechnical problems, but they also did so as a process of differentiating themselves as engineers. This brings us back to our initial concern with the co-production of bacteria, the field of synthetic biology and academic disciplinary identities, as we are now able to show how it was that the terminology, practices and promises of synthetic biology were brought into the local context of water engineering and innovation, and how our colleagues worked to make them fit their existing practices.

Biographies: narrating changes in co-productive relations

> I think probably the position I started from was that bacteria were a fairly small thing, physically small and small in terms of their role. I think I've changed. But that's where most engineers have classically come from, so most of my colleagues around here would have that viewpoint. Yet when you start looking at the diversity, what they're doing, what they can do, you see the kind of diversity you've got in a very physically small area is huge. And then considering the all-pervading nature of them, just every surface, everything around us will have this covering of microbial communities. So even with the water distribution system the classic engineer's view is that we take this surface water, with lots of organics in it, we put it through this great treatment chain, we've treated it to wonderfully high standards, and now we've got this beautiful resource. And we chuck it in the pipe and it magically comes out the other end the same standard. And to them these pipes are closed, there's no way in there, there are no bugs. Everyone sees the system as an inert, non-reactive system. I've very much got a different opinion of this. Water distribution systems are where a lot of our problems come from. Partly because of the huge surface areas of the pipes you've got in there that biofilms do grow on, are growing on. They're not big thick slimes you're going to see and worry about, but they are there, they're all pervasive. So I think it is probably the all-pervading nature of microbial communities and biofilms that's probably different [in how I think compared] to most engineers. (Academic Water Engineer 1)

In this quotation, we can see another narrative of bacterial emergence. Importantly, this example draws a close connection between how the engineer's thinking has changed and how he himself has changed, and relates this to the ways in which he is different to other engineering colleagues. This is a common theme in such narratives. In the previous section, for example, we described how a chemical engineering colleague had set about learning how to understand biology, and begun attending conferences, workshops and so on, in order to begin to integrate bacteria into his research in a way that better fitted their ontological enactment as lively and active. He worked to change himself and his practices so as to be better attuned to the agency of bacteria. His own agency and that of the bacteria were thus co-produced within the situation of an engineering problematisation ongoing in the professional life of the academic.

In this section we introduce the term '*bio*graphies' to refer to this kind of ontological work, regarding the ways in which academics, industrialists, policy-makers and other actors describe changes in their understandings of the ontological status of bio-objects over time, and how this has been entangled with various changes in their practices, in how they identify themselves as professionals and in how broader projects of knowledge-making, governance and social life are organised.

Scholarship that we have reviewed on bio-objectification so far has usefully drawn our attention to the co-production of bio-objects and sociopolitical contexts, in particular as regards to how the co-production of new bio-objects from generative relations often results in disruptions to those very relations and to classification systems. However, the literature on bio-objectification has paid too little attention to the ways in which other actors' identities are changed or disrupted as a result of this co-production. In developing the concept of biographies, we hope to remedy this. Our emphasis, given the context of our investigation into synthetic biology, will be on how ontological shifts in bacteria occurred through bio-objectification processes that were entangled with changes in disciplinary identity and the construction of a new sociotechnical field. Importantly, we also position biographies as a term to operate at the level of everyday academic life. However, we mean for it to be useful across a range of other contexts in which many different kinds of bio-objects are situated, and in which the relation between human actors and those bio-objects is co-produced in some way, not necessarily around shifts in academic practices.

In choosing the term biography, which is used generally to refer to the life of a person, we obviously give more prominence to the work of the human actors in this co-production. We do not mean to imply that non-human actors do not contribute to the co-production of bio-objects and biographies. Indeed, it is worth occasionally italicising the *bio* of biographies to remind us of the multiple living agents implied by the term. We have so far shown some of how bacteria express agency as they are enacted within sociotechnical practices. Moreover, in using the term biographies we also emphasise the human aspects of this co-production to point towards a second dimension of the concept, namely the ways in which these co-productive relations are themselves *narrated* as part of bio-objectification processes, shifts in governance regimes, articulations of disciplinarity, social science scholarship and so forth. In this regard, we are interested in how such descriptions of co-production are themselves produced by actors in strategic and political ways. This gives the term a temporal and reflexive dimension.

An important feature of the enactment of bio-objects is that they will always be situated within a particular sociotechnical space. This helps to account for how bio-objects change across space and time. It also invites us to be cognisant of how actors' descriptions of these processes will change through the accumulation and loss, over time, of particular social ties, forms of capital and embodied capacities and knowledge. We have so far described some of this work in our examination of actors' narratives of bacterial emergence, and the sociopolitical context of water governance and industrial innovation in which this emergence of bacteria is ongoing. Moreover, biographies will also always be situated in terms of power dynamics, strategic games and larger life projects.

By the time that we began our ethnography of everyday work in the project, a select number of the molecular biologists and chemical engineers involved had already begun to call themselves, if only sometimes, synthetic biologists. A range of factors were important to them being able sometimes to claim this shift in professional identification. Most obviously, these particular individuals had been central to winning a grant for one of the UK synthetic biology networks, funded by RCUK, and so were involved in consolidating the synthetic biology community in the UK at an institutional level. They also regularly attended synthetic biology workshops and conferences, such as the SBX.0, they were in the process of setting-up a synthetic biology Masters programme and tended more towards the use of synthetic biology terminologies over those of molecular biology in their everyday talk. Moreover, they had begun concertedly to reorganise some of their research practices, both in terms of laboratory or computational work and intellectual framing, so that they better fitted the emerging norms of the broader community of academics calling themselves synthetic biologists. They had started using some standard parts and the Registry, and they were thinking about how to optimise and characterise their parts and chassis more effectively. However, even these actors' use of the term synthetic biology changed in how they identified it within their research themes, everyday experimental practices, career trajectories, disciplinary identities and so on across different situations and over time.

One of the most senior scientists in the project, who perhaps used the term more frequently than anyone else in the group, made very strategic use of the ideas and practices of synthetic biology within different dimensions of his everyday working life. From his work prior to the emergence of synthetic biology to his contemporary role in driving forward an agenda informed by synthetic biology, his strategic use of the field is best understood within his everyday academic practices, responsibilities and ambitions. As such, his narratives and our observations of his work are worth examining in some detail in order to flesh out the various connections in bio-objectification, sociotechnical politics and *bio*graphy that we have so far developed and to evidence how they can be applied together to understand changing practices around synthetic biology.

Although he regularly referred to himself as someone 'doing synthetic biology', in interviews Academic Chemical Engineer 1 tended to refer to himself as a 'chemical engineer' rather than as a synthetic biologist. He also connected this question of disciplinary identity to the importance of learning interdisciplinary practices in synthetic biology:

> Academic Chemical Engineer 1: [Synthetic biology] requires, well certainly if we just talk about some of the core things, it requires an engineering mind-set, but it requires, it should require an understanding of the biology, and so those two things are often not completely connected together with every person. And so I think that just to do the work itself, in the lab, it certainly requires an understanding of an engineering mind-set and the technical skills of being able to do the biology. So that's just one example of why synthetic biology is an engineering discipline – is multi-disciplinary.

> Andy Balmer: Is synthetic biology a discipline in its own right?

> ACE 1: No, not yet. It's multi-disciplinary. So it's an area. But I'm not sure it's a discipline yet. Because I mean, well yeah, do I view myself as a synthetic biologist? No, I view myself as a chemical engineer.

In this quotation our colleague linked the question of interdisciplinarity in synthetic biology to individual practical understanding, arguing that synthetic biology's interdisciplinary demand had to be embodied within the individual scientist, rather than being simply the bringing together of engineers with biologists. He saw synthetic biology as a practice that had become available for engineers to use but not as a discipline in its own right. The people working in his group also tended not to describe themselves as synthetic biologists but rather as doing synthetic biology. At this stage, at least in our situation, the field was something that people used or did, not something that they became.

At the same time, in describing his identity as a chemical engineer, our colleague nonetheless differentiated himself from other chemical engineers through his narrative of bacterial emergence. He did so by stressing how his use of the engineering perspective on biological phenomena made him different to other engineers. However, he also distinguished himself from biologists and molecular geneticists. This came out quite clearly in how he described his first engagements with biologists:

> Certainly in my discussions with biologists, what confused me a lot of the time, when I was first trying to get into molecular biology in particular, was that every biologist that I met at the time had their favourite organism. That was their favourite organism so they studied it, not because it was the best organism for doing that function, for making hydrogen for example. And I just didn't understand why you'd do that. At the time they'd never even compared them, they'd never taken two organisms and asked does organism A make more hydrogen that organism B. So things like that were quite odd as an engineer. Because if you were looking at catalysts or widgets or cars that go faster you'd line them up and see which went faster, but those things just didn't occur to people. I'm not saying I'm the only one to have noticed but in the particular sphere I was working in, I was a bit baffled. (Academic Chemical Engineer 1)

In this quotation, in the midst of first developing his expertise in biological phenomena, our engineering collaborator distinguishes himself from biologists by reference to his disciplinary identity 'as an engineer.' Such narratives of engineers' distinction from other engineers and from biologists are common both at the core of synthetic biology as well as in our actors' accounts in this project. As Tom Knight, so-called 'father' of synthetic biology (Coghlan, 2012) says:

> A biologist goes into the lab, studies a system and finds that it is far more complex than anyone suspected. He's delighted; he can spend a lot of time exploring that complexity and writing papers about it. An engineer goes into the lab and makes the same finding. His response is: 'How can I get rid of this?' (Quoted in Brown, 2004)

Such narratives form part of our colleagues' accounts of their professional trajectory. As we have pointed out already, these refer first to a shift in ontological co-production of bacteria prior to the emergence of synthetic biology. These earlier changes were crucial to how actors in our project were primed for the emerging set of values, practices, imaginaries and so on of synthetic biology as it was being developed at the core in opposition to molecular biology. Their narratives of bacterial emergence not only helped them to describe how they had come to work on biological phenomena in the way that they did, but also served the purposes of our actors in that they helped them to define themselves as being the kinds of people who would fit into synthetic biology because of their professional biographies.

Indeed, telling their stories of bacterial emergence was strategic in as much as it reflected the kinds of stories being told in the news and at conferences by more prominent scientists at the core of the field. Such descriptions of shifts in the ontology of bacteria and related epistemological practices were made as part of the construction of the field, in the local situation, in relation to the core and within a broader narrative of professional development and disciplinarity. Narrating bacteria and disciplines in such a way helped our colleagues to circumscribe a space between their former engineering identities whilst retaining distance from academics in biological disciplines. It was from this in-between space that engineers in the project sought to help to make synthetic biology at the university a reality for both their own research practices and professional development, and at the institution more broadly. Much as we saw in Chapter 2, what exactly would come to constitute synthetic biology in this situation, however, was not a straightforward transposition of practices from the core to the periphery. Instead, being primed for synthetic biology practices, and narrating their professional trajectories in this way, resulted in our actors having to negotiate various entanglements between emerging features of synthetic biology at the core and the extant features of the local situation.

Academic Chemical Engineer 1 had been one of the key actors in trying to bring synthetic biology into the university and in making the project we were working on possible. He had accomplished this by partially adopting the terminology of synthetic biology and by learning more about the core of the emerging field from within his own everyday practices as a chemical engineer working with biological phenomena.

> I suppose in 2000–2001, we got some money in areas that weren't really synthetic biology but they were at the construction end of, it wasn't called synthetic biology then, it was more metabolic engineering, on trying to make biofuels-type devices I suppose, using

bacteria. And, it became more obvious as years went by that some of the things we were doing could be labelled synthetic biology, and so I started getting interested in synthetic biology and then got invited to be part of a consortium [on biofuels...]. So I just got much more immersed in the synthetic biology community. (Academic Chemical Engineer 1)

In this regard, longer-standing shifts in the co-production of the engineer's identity and practices of interacting with bacteria allowed him to describe some of his and his lab group's activities as synthetic biology as the field began to emerge more prominently out of the US and certain parts of the UK and Europe. His engagements then began to develop with core synthetic biology, through attending workshops and conferences, reading papers, attuning himself to the imaginaries of the field and so on. These were important factors in him helping to bring funding into the university to pursue research specifically under this banner. As he continued:

I was keen to see that the University had some presence in synthetic biology, because I don't think we've had a very big presence in systems biology, and I thought synthetic biology is very much an engineering-led discipline. ... So the idea of the BBSRC networks in synthetic biology came up and I pushed quite hard to make sure we got a network here. So we arranged lots of [local] networks, seminars and that kind of thing to pull the proposal together. That's how the network got funded. (Academic Chemical Engineer 1)

Part of the reason for wanting to work hard to get a good grant application together for the RCUK Networks in Synthetic Biology call related to the engineer's sense that he and others working in his department had missed an opportunity to get funding under the banner of systems biology. This was another important factor to assembling a network of researchers to bid for synthetic biology funds. In fact, other institutional factors also played a role in the development of synthetic biology in the specific form that it took at the university, which we will come to later.

At this point in our colleague's past, the RCUK had begun to push towards synthetic biology with more determination. A drive to ensure that the UK was not left behind in the global race to dominate the biotechnology industry had begun to place synthetic biology at the forefront of national narratives around the promised bioeconomy. The first major RCUK investment was to begin actively organising networks of scientists and engineers into teams that would workshop ideas, go to conferences together, network with industry and so on in order to then bid for larger pots of money in future synthetic biology and industrial biotechnology calls (Molyneux-Hodgson and Meyer, 2009).

Having worked hard to establish relevant connections at the university, Academic Chemical Engineer 1 then worked with some of these actors to leverage their expertise into a call for funding from a separate RCUK funding pot.

> Obviously as part of that conversation we made various connections around the University. So when the funding call for this project came up we had the idea for it to be related to the water industry. [Academic Environmental Engineer 1] had already been aware of the growing synthetic biology initiatives in the University and in the department, so conversations started to arise then about [...] how to maximise the critical mass that we'd already started to get together with synthetic biology with the critical mass that they already had on the water side, and I guess it was the merging of two strengths, if you like, if you call synthetic biology a strength, it's not as an established strength as is the water engineering centre of the University, it's growth, it's something we're trying to do. (Academic Chemical Engineer 1)

The successful grant application for a network in synthetic biology at the university was leveraged as capital in the application for funding to study the prospects of the field for the water industry. As our colleague described, this was partly because the university already had significant expertise and experimental infrastructure for the study of water engineering. So these established research interests, groups of people, laboratories and so forth were enrolled in order to bid for synthetic biology funding to work on long-standing concerns in the water research group. However, it was as much about keeping the ball rolling on synthetic biology as it was about continuing the critical mass on water research. How synthetic biology was enacted at the university thus depended upon the kinds of opportunities that came up in the RCUK, on existing relations and networks, on what could be pieced together quickly and on extant expertise and centres of academic capital.

Using the terminology and practices of synthetic biology, and indeed calling our project a synthetic biology project, was a strategic manoeuvre, situated within the professional biographies of the academics working in the university. There were clear ways in which the disciplinary biographies of the academics in the project were being developed through applying for this kind of funding. It helped them to claim more substantively that synthetic biology, as found at the core, was something they were 'trying to do.' However, they were also trying to make it work within their particular circumstances. Describing and orienting to bacteria in particular ways through adopting the language and practices of synthetic biology helped our colleagues to place themselves strategically within the university, to stick themselves to existing research clusters and to progress their own careers.

For example, synthetic biology was made to fit into a larger set of institutional politics. The academics working in engineering departments had begun to work on biological problems and had done so by differentiating their approach not only from other engineers in their discipline but also from biologists and molecular geneticists in the life sciences faculty. In part this was an epistemological concern, but it was also very much about the local professional politics at the institution since colleagues in the biological disciplines had – not uniformly, but quite consistently – so far refused to engage with engineers on biological problems. Despite serious investments in large pieces of laboratory equipment in the engineering department, equipment not available at that time in the biological sciences building, there was little engagement from the biologists.

In a sense this was a kind of elitism in the biological disciplines. Or at least that was how it was understood by our engineers, and there was some evidence for this. One of the only molecular biologists involved with the synthetic biology project routinely complained about the lack of normalised molecular biology practices in the engineering labs. He rejected their laboratory and he largely rejected the synthetic biology practices that colleagues were trying to develop. Whilst contributing to the supervision of the iGEM team, he trained them to conduct genetic manipulations through his own routine practices rather than by using the techniques prescribed by iGEM's standards and protocols. Of course, this was not simply about elitism, as we have described elsewhere (Balmer and Bulpin, 2013) but also about saving time in a very pressured situation, which we explore further in Chapter 4. For now, we can say that it does at least point to the ways in which academics in the biological sciences ignored or resisted the emerging practices of synthetic biology within the engineering buildings.

There was also a sense of animosity from molecular biologists in the life sciences, who were not directly involved in this project, regarding the funding that the engineers had received from RCUK and from the university to remodel the engineering labs, purchase molecular technologies and equipment and to employ staff in these areas. In effect, there was a small-scale turf war going on. Synthetic biology represented for our engineers not just a set of practices and terms for engineering biology that sat more comfortably with their existing practices but also a way of differentiating the work that they were doing from work being done in the biological sciences, as part of an ongoing tussle regarding who should be supported to be doing this kind of research with microorganisms.

This all meant that molecular biologists from the biological sciences departments were noticeable by their absence in the synthetic biology networks, save for a select few individuals. So the networks in synthetic biology being constructed at the university, and elsewhere, were largely populated by engineers of various kinds. As such, when the funding applications were being constructed it was around problems being worked on in other departments at the university, for example in medicine and water engineering, outside of the life sciences departments. The engineers' strategy in setting up these networks was thus to accrue intellectual, practical and institutional capital over time, to make synthetic biology happen in their own professional biographies and in the engineering department by tying themselves into existing research units outside of the biological sciences, but in which biological phenomena were significant. Synthetic biology was spreading around the university through making particular social ties and material connections, and by linking synthetic biology with ongoing problems in other disciplines where the promises of control, standardisation, industrialisation and so forth might take root.

The development of a series of economic and sociotechnical promissory narratives in synthetic biology lent themselves to adoption in water engineering and the water industry, for example, because of the complementary concerns regarding innovation barriers, global challenges and so on that we described in Chapter 2. However, it was also hoped that the epistemology of synthetic biology – as it was being fitted together in the local situation, with the emphasis on control of complexity, optimisation of processes and overt adoption of engineering language and practices – might prove particularly useful in this area. This was because of the existing movements being made to shift industrial conceptualisations of bacteria and their biovalue within the massive assemblage of pipes, treatment facilities, waste and so forth.

Academic engineers had long been working to get industrial actors to reconstruct their ontology of bacteria in a similar fashion to their own narratives of bacterial emergence. They had been doing so since they had hit walls in their research because the ontology of bacteria as being inert was no longer sustainable. In order to move forward in their engineering work they had to begin to enact bacteria in new ways, although this was also risky business. As we will describe more fully in the next chapter, bacteria were positioned in a particular way in the industry because of a particular public health narrative on which the industry is founded and on which it makes its profits. Bacterial emergence in the water industry was ambiguous. Bacteria had begun to represent both a resource and a threat as they emerged in a more lively fashion within engineering problems like sewer sedimentation.

So our colleagues made strategic use of synthetic biology terminology and practices not only to be able to secure new sources of funding for their work and to make progress in their turf war at the university but also to address this ambiguity in the ontology of bacteria as they emerged in the industry. This was manifest during the industry open days, as we described in the previous chapter, during which engineers on the project worked to 'sell' synthetic biology to the industrial actors. They did this by describing bacteria in particular ways using the terms of synthetic biology, by emphasising engineering practices, by stressing how this was an exciting new field in which there would be lots of public funding coming up and by calling themselves – on choice occasions – synthetic biologists.

Examining these mundane, everyday features of the emergence of synthetic biology within boring problems like sewer sedimentation, biographies of academic careers and institutional politics helps to explain how it is that a novel, high-tech field like synthetic biology makes its way into the world. It was through engineers doing everyday things, trying to progress their research, their field, their careers and their departments that synthetic biology was made, and the ontologies of bacteria changed.

Conclusion

Actors at the core of synthetic biology have put forward particular promises regarding the field's economic potential, as we explored in the previous chapter. Here we have examined how such promises hinge on a particular ontology of bacteria, as well as other biological phenomena, and a set of practices designed to enact them as such. In this regard, synthetic biologists want to change what bacteria are.

In order to understand how ontologies of bacteria change as the field's terminology and practices are concretised, adopted, transformed and so forth, we have shown that it is helpful to look closely at the history, politics, power relations and everyday research activities in which such changes are being negotiated. In short, synthetic biology as a field and its bio-objects are generated through everyday relations. What bacteria are, how they are understood and how they are controlled in relation to synthetic biology are best articulated at the level of the co-production of bio-objects, disciplines and identities within a specific situation.

We have shown that the kinds of transformations in bio-objects that are being enacted in the core do not carry over without change into local situations, in this context in the periphery of the field and the water industry. Instead, the practices of synthetic biology are being integrated into longer-term projects of academic research. Changes in the ontological status of bio-objects like bacteria, for example during 'bacterial emergence', are not only tied to the development of novel epistemic practices and fields, like those of synthetic biology, but also to the development of a professional trajectory. And, as we have explored, professional trajectories are deeply personal features of academic's lives, which weave in and through the everyday practices of scientific research. In this regard, the ontology of bacteria in synthetic biology practices and the field's sociopolitical construction is very much also a personal matter. It is *matter* in part transformed by the personal lives of scientists. At the same time, the materiality of bacteria helped to bring about changes in academic's lives and their practices, since they are wilful. They resisted certain practices, for example in being modelled as inert, whilst they helped to enact others, for example in shaping how sediment could be conceptualised and studied.

Moreover, synthetic biology is being organised in relation to broader reconfigurations of academic life. We saw this in the previous chapter, as regards global challenges and the 'valley of death' and how these impacted on the links between industry and academia. In this chapter, we have seen how synthetic biology was enacted through the reconfiguration of an engineering department, which was intended to help ensure that the department remained competitive in funding bids to the RCUK and other organisations, and was able to bid for emerging money in the context of industrial biotechnology. The development of new ways of working with bacteria is thus entangled with governance.

We have evidenced a deeply co-productive relationship: changes in synthetic biology help to bring about changes in governance, which help to bring about changes in universities and academic careers, which help to bring about changes in synthetic biology. They are mutually constitutive at a number of different levels, and all of this helps to enact microorganisms in particular ways.

Overall, then, the emergence of synthetic biology, with all of its epistemological and ontological implications, is being negotiated both in mundane and routine practices of everyday academic practices and strategically as part of everyday professional lives.

Coda 3
Critics on the Inside

Andy Balmer: So, talking about social science and natural science, how do such disparate groups collaborate do you think?

Academic Environmental Engineer 1: I guess it's a little bit like coming up with an idea that you might not have come up with on your own. It doesn't necessarily matter, you know, who was the one who came up with it in the first place. The discussions and the interactions between opposite, more disparate disciplines mean that you can approach a problem in a slightly different way, or get a slightly different angle on it. Together you come up with an idea that is better than if you'd just sat there on your own. My interactions with the social sciences, I think, are a very exciting thing. You guys are making me think things in a slightly different way, and it is better because of it.

In our sociological research and attempts at collaboration we tried to abstain from defining synthetic biology and instead worked to understand how the question of what was, or should be, constitutive of synthetic biology was dealt with within the project itself. Our abstention was methodologically useful, and to some degree there was an implicit politics in this disposition. The maintenance of ambiguity around what synthetic biology is and how it should be enacted has been a more or less explicit political project in STS so far. The aim has been to keep synthetic biology open in the hope of helping to bring about a more democratic and plural space by encouraging our technical colleagues to engage more reflexively in how they constitute the field.

However, this approach became disruptive in our efforts to work with engineering colleagues to explore the possibility of using synthetic biology in the water industry. There were a number of ways in which we became inadvertently disruptive, and others in which we purposefully played with and worked towards disruption in one form or another. As regards purposeful disruption, there is a tradition in STS, having to do with ethnomethodological 'breaching experiments', reflexive games, opening up controversies and so on (Ashmore, 1989) in which disruptions can be understood to be productive. Moreover, as part of a broader community of STS scholars in the UK, we were committed to finding new ways to engage with colleagues in the natural and social sciences that were more 'experimental' (Balmer *et al.*, 2012).

Being on the inside of the project, whether as a co-investigator, post-doctoral research associate or a PhD student exploring a case study, we each provided critical disruption. By this we mean that from within the everyday practices of writing grants and papers, attending meetings and conferences and doing computational or laboratory work, we asked questions and made comments or suggestions using critical apparatus from STS and related fields. For example, throughout the project we actively sought to reflect the emerging analysis of synthetic biology in the STS literature. We gave talks in which we discussed promises and expectations; during meetings we would explore how different regimes of sharing, patenting and so forth were being invoked or not, and at conferences we chatted about how actors from the core were working to make synthetic biology in particular ways. In the lab we drew attention to tacit knowledge, shifting practices, and the creation of bio-objects, as well as raising or analysing a host of other situated topics or objects of interest as we went about our daily interactions. Although we talk about these kinds of engagements as being 'critical' they were not generally 'negative' in the everyday sense of the term. Instead, these conversations and presentations mostly had a positive and engaging dynamic. Our colleagues in the natural sciences were often eager to discuss afterwards what had been going on in the subtext of a meeting, to talk, gossip or laugh together at a conference or to find examples from their own life of topics we had presented in the lab group seminars. Our critical disruptions, then, were often received and responded to with enthusiasm and hospitality, and we tried intentionally to contribute to this atmosphere of collaboration.

As regards the more inadvertent effects of our critical disruptions, however, we want to focus particularly on the effects of our persistent ambivalence to synthetic biology. It is worth looking at a few examples in which our intentional and unintentional disruptions got murky, messy or entangled. First, as we have described in the previous chapters, our colleagues were emphatically trying to make synthetic biology useful to industrial actors and to 'sell' the idea of synthetic biology to them. A number of engineers in the project understood this to be our role. They sometimes saw us to be experts in public and industry relations, a special kind of academic PR. As such, we were occasionally asked for information on how they could best market synthetic biology to the industry, on how to frame it and what to avoid saying. We were keen not to position ourselves as marketing or PR experts, and instead we sought to challenge, and so to intentionally disrupt, how various notions of the public, industry, innovation and regulation were being constructed and used. We also made it clear that we were not so much interested in how to promote synthetic biology, but to understand how this work was happening and what 'synthetic biology' was or could be in this situation. In this regard, we sought to be constructive, by use of such critical tools. We wanted to help to make synthetic biology at the university in ways that were more reflexive, politically minded, engaged with external actors and so forth. Our critical disruptions were also constructions. This effort was reasonably successful and a number of our collaborators began at least to understand what it was that we were trying to do, and began less explicitly to seek advice on the issue of PR. Moreover, some of them began using our critical engagement and ambivalence to synthetic biology in their own formulations of the field. As one colleague put it, 'You guys are making me think things in a slightly different way, and it is better because of it' (Academic Environmental Engineer 1).

Others, however, on realising that we did not plan on greasing the wheels of synthetic biology's journey into the water industry, simply disengaged from our work and paid negligible interest to what we were doing, or made only the most minimal effort to work with us. These actors had mostly been interested in how they might overcome the 'social barriers' to innovation, and were perhaps less engaged with synthetic biology more generally. Once they had lost interest in how we might help them to do this, they also seemed to lose interest in synthetic biology and the project, to some extent. In this way, our presence and critical disposition towards synthetic biology contributed to shaping how actors in the project interacted, along with a number of other factors, so that those whom became less involved with us were also the ones whom became less involved in the project more generally. Our intentional disruptions of concepts and practices, and less intentional disruptions of the team's structure and working relationships were thus connected. What we were trying to disrupt and construct, exactly, also became a bit messy, and it was sometimes hard to know how our interventions represented attempts at collaboration and making something together, and not just STS as standard.

Second, in trying to explore with our colleagues some of the ways in which they made sense of synthetic biology, we regularly asked questions that prompted them to account for their usage of the term. During our interviews we asked them to tell us what they thought synthetic biology was. During participant observations we would also explore how the research related to synthetic biology by asking about how what was happening was different or similar to their more long-standing practices. This was often understood as a challenge rather than as a more dispassionate inquiry. As a consequence, we found ourselves in a number of situations in which our questions implicitly challenged our colleagues' use of the term or their claims to having been engaged in synthetic biology practices. Sometimes it clearly seemed to them that we would not let the matter rest. One particular instance is exemplary and worth exploring further, since it also points to some more resources and strategies used by the engineers to define their work as synthetic biology and to how they themselves questioned the specificity of the term and what the field was becoming.

About two-thirds of the way through the project's timeline, a meeting was convened, which had a dual agenda: 1) for Andy to ask questions about how a couple of new people had been added to the project; and 2) to determine how an experiment should be organised, designed and implemented. In the midst of this meeting, which moved back and forth between the two subjects, the meaning of synthetic biology and how it related to the activity of the project arose again. Andy had been asking about what had been done so far in the grant and why certain people had been chosen to conduct particular tasks or had been brought into particular projects. The answers seemed to do with the need for their expertise; their current status of employment; the loss of people to other projects, labs or institutions and their closeness to other actors in the team. As part of these discussions Andy ended up summarising some of what had been done to date in the project, and then posed a reflection regarding how much of what had happened so far seemed to be quite familiar to him as quite well-established practices of genetic engineering. As we have described, a number of the team members had begun using the language of synthetic biology, for example in choosing to say 'parts', 'devices' and 'chassis' where they might previously or otherwise have used 'genes', 'construct' or 'bugs'. And so Andy posed the question of how these elements were represented in the project, 'Where are the parts and devices that you're talking about?'

Andy's intervention in the conversation had the effect of challenging whether the terminology being used accurately reflected the practices and objects used in the laboratories. Perhaps his questions also embedded an unvoiced scepticism about whether synthetic biology was anything new, at least in how it was then being configured. This at least blurred the boundary between ambivalence and doubt, since the question asked our colleagues to link more explicitly terminological changes to changes in their practices and associated materials. There were indeed some ways in which such shifts could be seen, and we had been charting some of them ourselves. However, we also wanted to understand how the engineering academics negotiated and articulated these changes. Academic Environmental Engineer 1 responded to the question by listing a series of ways in which their project was connected to the ethos and methods of synthetic biology. She described how they had produced some data that would be of use to synthetic biologists because it would help to characterise certain parts, that they had produced a list of parts in registries that they intended to use in their next experiment and that the iGEM team had submitted parts to the BioBricks™ library. She added that in the most immediately forthcoming experiments they were going to be working with bacteria sourced from another lab and that they intended to further modify these organisms to suit their own purposes so as to optimise them for their specific context. Andy suggested that much of this could be viewed as traditional genetic engineering, even if, on some accounts, that might be an uncharitable reading.

Discussion then moved towards the design of the experiment, which was, in part, to ease a little of the tension that had built up. Meanwhile, the project lead rooted through some files at the back of the room. She returned with a copy of the Royal Academy of Engineering's report on synthetic biology, at which point a further series of exchanges about the definition of the field and of the project began:

Academic Environmental Engineer 1: According to the Royal Academy of Engineering's definition, which I will use to defend our project -

Andy Balmer: And which you've literally got out [laughing].

AEE 1: [laughing] I had to, to make sure! Now that I know I'm being taped [gesturing to the Dictaphone, and then reading from the report]: 'Synthetic biology aims to design and engineer biologically based parts,' which *we're* [indicating the academics, not the iGEM team] not really doing just yet, 'as well as redesigning existing natural systems,' which we are already and will continue to be doing.

AB: So why is it that you've got the report out? Why has my asking if this is a synthetic biology project and how you're using engineering prompted you to defend that it is synthetic biology in this way?

AEE 1: Well because when [Academic Molecular Biologist 1] started on the project, looking at the pathogen detection, I kept asking myself how is this synthetic biology and not just molecular biology or standard biology? So I get nervous that the [funding body] is going to ask this question. I think it is synthetic biology because we are [redesigning the organism] to have some sort of an engineering advantage. So for me that's synthetic biology.

The engineer's response used the definitional power of the institutional authority, The Royal Academy of Engineering, to secure the project's status as falling within synthetic biology practices. The literal text of the report, wielded in this conversation, embedded the institution's authority in a material form, that can be – and in this context was – used to authorise the terminology 'synthetic biology' in local situations. In this respect, the disciplinary identity of the project and the nomenclature of the research was made more concrete by being tied to the academy. The text acted as a mechanism for linking together the core and the periphery.

However, another engineering authority, the funding body, was subsequently invoked as a threat since its effects were more keenly felt in everyday academic life. The institutional practices of peer review and project evaluation were connected, in talk that followed the above quotations, to the anxiety about the definition of synthetic biology. Specifically, a discussion began about whether prominent synthetic biologists were, at that time, part of the funding body's review panels and whether – if faced with evaluating this project and future projects – they would accept this definition of synthetic biology and legitimate its use in describing the research. In our presence, and by virtue of the ambivalence or scepticism that we brought into our research practices, disruption was caused by opening up the situated meaning of synthetic biology to discussion, and thereby opening up a series of power relations between individual scientists, institutional authorities and disciplines.

In both of these instances our presence as sociologists, with particular methods and particular concerns, opened up or disrupted the research practices and adoption of terminologies that our colleagues were trying to implement. We had become critics on the inside of the project in ways that were sometimes inadvertently and unproductively disruptive. As the project further developed after this incident, it became much more difficult, from an interpersonal standpoint, to ask questions explicitly regarding the status of synthetic biology. If the use of the Royal Academy Report had not been completely successful in closing down ambiguities around synthetic biology in the local situation, it had more forcefully had the effect of signifying our colleagues' frustrations with our desire to maintain those ambiguities. In this regard, some of the 'subterranean logics of ambivalence, reserve and critique,' (Fitzgerald *et al.*, 2014b) had surfaced. To some extent then, we now also had to become critics on the inside in a more corporeal sense. We had to keep silent. Keep our thoughts and language to ourselves.

Having contributed to this disruption we had perhaps also come to signify the difficulties that our colleagues were having in engineering biological parts and in getting the industry on board with synthetic biology. Indeed, there was sometimes the sense, and on one or two occasions it became explicit, that the money spent on employing social scientists in the grant would have been better used to further the immediate scientific aims of engineering novel microorganisms.

Providing critique was part of how we did our sociological practices within the project team's attempts to collaborate. There were ways in which this was productive or constructive, and ways in which it became less so. What is clear, however, is that the value of social science in this situation was very much posed through the larger problematisation with which our colleagues were concerned, namely to introduce synthetic biology into the water industry and so to demonstrate the field's value in less obvious venues. In some ways we contributed to this work by helping to reconfigure understandings of the problem of industrial and academic innovation and of how the water industry worked. In some ways we stifled this work, by less intentionally disrupting progress in this direction by keeping open some issues that our colleagues had tried to shut down. Our interest in the sociopolitical dimensions of changing practices in the field of synthetic biology therefore brought politics too prominently into the lab and into meetings for some of the engineers.

There remains a question then, in contemporary STS entanglements, regarding the extent to which a balance should or could be struck between opening up and closing down, between disrupting and constructing sociotechnical concepts, objects and practices as part of collaborations. The default STS position is to keep things open since we are acutely aware of the default science position to close things down. Because doors are shutting all around it can become habitual to try to open windows. However, there might be points at which social scientists are confronted with the question of whether keeping things open entails a form of disruption that cannot be sustained, without undoing the practices in which we are mutually entangled. To what extent do we have to take responsibility for our contributions to the success or failure of scientific and industrial innovation practices in which we collaborate?

4
Bodies

A robotic arm moves silently along a lab bench carrying a 96-well plate. Another automated device pipettes liquid into the wells before the plate is smoothly transported by a second robotic arm and deposited in a microplate reader. The plate briefly disappears inside the opaque, grey box before the cycle of activity continues.

Introduction

This robotic sequence plays out in a video on the website of Ginkgo Bioworks, a synthetic biology start-up company located in Boston with the explicit mission to 'make biology easier to engineer' (Ginkgo, 2015a). Ginkgo has become emblematic of the promised industrial future of synthetic biology, and it is certainly a core enterprise. One of the principle ways in which Ginkgo seeks to realise synthetic biology's visions is by replacing traditional scientific practices of working 'by hand' (Ginkgo, 2015b) with automated, robotic processes. In this regard, the vision of industrialisation is set up in relation to the production of 'readymade' bio-objects, procedures and tools. Biological spaces, labour and methods are being reconfigured to align with industrial aspirations and, in keeping with this, the Ginkgo laboratory has been renamed a 'foundry', in which automated mechanisms construct such readymades. In new bio-industrial settings like these, the hands and bodies of scientists seem conspicuous by their absence.

Most academic synthetic biology projects, however, are developing from within the routines and long-standing embodied practices of laboratories, departments and research groups, as well as in relation to established problems and sites of industrial production, rather than in

116

the custom-built spaces of start-ups. In this chapter, we continue our exploration of how our colleagues in the project worked to try to make synthetic biology fit within the existing everyday practices of the university and those of the water industry. Our emphasis here however, is on how bodies figured, or not, as part of this work.

Where the body was once understood to be the preserve of the biological and medical sciences, it is now a central concern in a diverse range of sociological debates, and features prominently in a number of theoretical and empirical approaches. Sociological inquiry into the body has developed within, and informed research on, a vast array of issues including gender, ageing, race, ethnicity and health. Concepts of the body and experiences of embodiment are now routinely theorised as socially constructed and embedded within social practices (Crossley, 2001; Turner, 2008). In turn, the body is understood as central to understanding society, sociomaterial interactions and everyday life. As Turner (1992: 3) writes:

> In order to comprehend the everyday world, or life-world, it appears to me that a sociology of the body is a necessary condition for understanding everyday routines, conditions and requirements. Everyday life is about the production and reproduction of bodies[.]

For many sociologists, the body is understood to be a key site of political, cultural, social and economic intervention, becoming 'a contested terrain on which struggles over control and resistance are fought out in contemporary societies' (Hancock *et al.*, 2000: 1). Medical sociologists, feminist and science studies scholars have drawn on Foucauldian notions of biopower, knowledge and discipline to examine how technoscientific discourses contribute to the governance and 'medicalisation' of the body (Hughes, 2000; Lupton, 1997; Rose, 2007b; Sawicki, 1991). Sociological studies of scientific and medical work have emphasised the ontological multiplicity of the human body as it is differentially enacted across various epistemic and health practices (Cussins, 1996; Mol, 2002). Notions of 'posthuman' and 'cyborg' bodies have also been deployed to explore how scientific and technical developments are forming new kinds of hybrid ontologies that reconfigure and disrupt traditional boundaries between humans and technology (Halberstam and Livingston, 1995; Haraway, 1991).

STS has also produced a significant literature on the body, not as an object of epistemic inquiry or medical intervention, but with regards to the role of the scientist's body in the production of knowledge. Early studies of scientific practice highlighted the importance of tacit knowledge (Polanyi, 1967) to the successful accomplishment of scientific techniques (Collins, 1974; MacKenzie and Spinardi, 1995). In particular, laboratory ethnographies have illuminated the embodied 'artisanal' craft of biological work (Cambrosio and Keating, 1988) and the continuing importance of these skills in biological fields despite the increasing use of standardised procedures, tools and materials (Knorr Cetina, 1999; Jordan and Lynch, 1992) across the biological sciences. The scientist's body is thus understood to be an essential 'instrument of inquiry' in the lab (Knorr Cetina, 1999: 94) where practical competencies emerge from a 'lengthy personal struggle with materials' (Jordan and Lynch, 1992: 90).

Ethnographic studies have also become attuned to the role of the senses in scientific practice, for example Goodwin's (1995) study of 'seeing' in oceanographic research ships and Mody's (2005) work on hearing in surface science laboratories. Myers and Dumit (2011) have begun to explore scientists' bodily movements and 'affective entanglements' with the objects and instruments of inquiry in experimental work.

Reflecting on the wealth of STS studies of scientific practices, Mody (2005: 176) writes that:

> We can discern the whole physical presence of laboratory workers...how they comport themselves, how they inhabit specially constructed lab spaces, how they interact with instruments and artifacts, how they shape and move their bodies to be perceived and disciplined by the gaze of others, and how their bodily experiences (their illnesses and exertions) are insinuated into their craft.

We are also mindful, however, that not all bodies are treated equally and that some kinds of labour are often left absent from accounts of scientific work (Shapin, 1989).

In reviewing such studies, Bulpin and Molyneux-Hodgson (2013) argue that the institutional routines and the discursive and material practices of particular epistemic settings can be understood as 'disciplining technologies' that constitute both the subjects and objects of scientific work. The everyday practices through which scientific subjects and scientific knowledge are co-produced are connected to the institutional organisation of disciplines through the organisation of scientists' bodies (Lenoir, 1997).

These literatures have produced important insights into the central role of the body in the production of scientific knowledge and how particular values, norms, skills and knowledge are embodied as part of a scientists' professional self. However, the rich descriptions of tacit knowledge and embodied skills found in these studies are primarily bound within the immediate epistemic setting of the laboratory space and tend to focus narrowly on the accomplishment of specific techniques. There is little examination of how the bodily ordering of technoscientific work is situated within broader sets of practices, 'transepistemic arenas' (Knorr Cetina, 1982) or socioeconomic systems. Furthermore, these studies have been made in established disciplinary contexts in which bodily actions have become routinised within relatively stable assemblages of places, objects, people and practices.

In this chapter, we add to STS accounts of the body and its relation to knowledge production by examining how scientists and engineers' bodies are problematised and configured in the context of changing epistemic practices in an emerging technoscience. We use the concept of the body as a 'hazard' or 'at hazard' to emphasise the role the body plays in the risky environment of changing practices. Conceptualising the body in this way draws attention to how the threat, and management, of failure permeates relations between epistemic practices, objects and subjects, and how these are organised in the everyday life of scientists and engineers. In the context of our project, we explore how new hazards might be created in shifting relations between the body, bio-objects and knowledge, from both within and beyond the immediate confines of the lab, both as practices shift and in the movement from the core to the periphery.

We investigate two contexts, the laboratory in which can be found scientific bodies, those of both novices and professionals, and the industrial and public context in which can be found those of process engineers, consumers' bodies and the 'public body' writ large. In each of these situations, there were long-standing ways in which the body featured as part of everyday practices and which were organised in relation to particular norms, politics and materialities. We call on the notion of 'hazards' in this investigation to help to articulate the entanglement of bodies and bacteria. First, by looking at how the body becomes a hazard to bacteria in the laboratory; and second, by looking at how bio-objects like bacteria become hazardous to the body in the context of the water industry. This lens helps us to understand how it is through everyday life and mundane situations that practices have to be changed in the context of synthetic biology. We begin by examining in more detail what proponents of synthetic biology have in mind for the body, before then moving into our two empirical sites: the iGEM competition laboratory, in which our novice team was competing to engineer a microorganism over the summer, and water treatment facilities, in which thousands of tons of water must be processed every day for input into the potable water system, or for return to the environment.

The body in synthetic biology

Synthetic biologists commonly distinguish their work from past biotechnological disciplines by contrasting their approach with the 'ad hoc' and 'artisanal' nature of traditional bioengineering practices (Shetty, 2008: 3). This helps to constitute synthetic biology as an engineering discipline with significant industrial promise. It also shifts the emphasis away from a creative engagement with materiality towards a more systematised encounter. Existing laboratory practices are understood to be unnecessarily time-consuming, relying upon lengthy apprenticeships to learn the uncodified, embodied skills required to manipulate biological matter in the lab successfully. The efforts of synthetic biologists to remake biological material into modular, standardised parts are intended to reduce reliance on such specialised craft skills with the ultimate ambition of designing processes for the automatic assembly of genetic parts. Synthetic biologists seek not only to 'get rid of' biological complexity through applying engineering principles and practices to biology but also to get rid of the skilled bodies of professional biologists in the process.

The drive for standardisation in synthetic biology is evident in the narratives of engineers' embodied experiences of being in molecular biology labs and the frustrations they felt with the localised, tacit knowledge needed to work with the unpredictable heterogeneity of biological material in such situations. Tom Knight, one of the 'founding fathers' of synthetic biology (and co-founder of Ginkgo Bioworks) describes how:

> biologists talk about having "good hands" in the lab, and everyone does experiments in different ways. [but] When you put two pieces of DNA together you need a standard way of doing this. (Tom Knight, quoted in Robbins, 2009: 9)

We came across this account of biology in our project, too, as engineering colleagues often described their bemusement with the emphasis placed by biologists on their personal relationship with microorganisms. As we described in Chapter 3, this was a part of how they differentiated their engineering approach, and also their own professional trajectories, from those common to bioscientists of various sorts.

Designing materials, tools, practices and procedures to displace and replace the human body is certainly neither a new phenomenon in biological fields nor in industrial arenas more broadly. The introduction of mechanised looms in the textile industry is a well-known example from the history books, and standardised technologies that automate a range of well-established laboratory practices are now commonplace in biological workspaces (Jordan and Lynch, 1998). Laboratory tools have made particular forms of tacit knowledge explicit, for example, standardised electrophoresis equipment replaced the need for 'golden hands' to separate DNA lengths using a centrifuge (Fujimura, 1988: 267). The disembodiment of scientific practice is linked to the introduction of technologies that 'outperform' bodily functions as well as to scientific distrust in the accuracy and reliability of the human senses (Knorr Cetina, 1999: 94).

Thus, we might understand the rhetoric of standardisation and automation in synthetic biology as merely an extension of a long-standing disembodying narrative in biotechnological fields. Describing such automated visions of biological work as a continuation of traditional bioengineering logics serves to unsettle definitions of synthetic biology as something distinctly new. In the promotional texts and videos on the Gingko Bioworks website, the lingering close-ups of automated machinery are key to marketing synthetic biology as a novel, cutting-edge technoscientific field able to realise the industrial potential and biovalue that past biotechnological endeavours are seen to have failed to deliver. It recalls the now commonplace robotised production line of automated car manufacture and the reliability, safety and familiarity that this captures. The synthetic biology experts who we observe in the promotional video are primarily shown outside of the lab environs, in the uncluttered spaces of meeting rooms conversing with each other and working on laptops.

Amidst the scenes of near human-free facilities, scientists can still sometimes be observed manipulating tools and materials inside laboratories. Certainly, in the sociological investigations of our project, actors' bodies were not only observed across the different academic and industrial spaces in which we engaged but also remained central to how practitioners understood and enacted synthetic biology practices. It was clear that the vision of automation put forward at the core of synthetic biology did not easily translate into the everyday working practices here in the periphery. In the sections that follow, we explore how synthetic biology's vision for the body figured in the local situation. Our colleagues in the project tried to represent the university as a place in which synthetic biology was happening and to evidence the field's value to the water industry. In doing so, they encountered and negotiated competing relations between the body, knowledge and bio-objects.

The International Genetically Engineered Machine (iGEM) competition

One of the ways in which advocates at the core of synthetic biology are attempting to build and nurture the new field is by enrolling undergraduate students in the iGEM competition. Each year registered teams are given a 'kit' containing some of the most commonly used or best functioning BioBricks™. These can also be ordered from DNA synthesis companies or made 'in house' using the information available on the Registry of Standard Biological Parts. Any new parts designed by the iGEM teams are then submitted to the Registry for future users to access.

When Drew Endy, an engineer with postdoctoral training and experience in genetics and microbiology went to the Massachussetts Institute of Technology (MIT) in 2002, he joined two prominent engineers – Tom Knight and Randy Rettberg – in co-founding the MIT Synthetic Biology Working Group and the Registry of Standard Biological Parts. Together, these engineers nurtured this nascent engineering field at MIT, starting a synthetic biology summer class in 2003. The next year, the course had expanded into a national, albeit small, contest with teams from six US universities competing. In 2005, it became 'international', thus establishing itself as the '*i*GEM' competition, and, by 2015, 281 teams were competing from around the world, although the distribution of competitors is weighted towards particular regions within Europe, Asia and the Americas. As the competition has expanded, so too have the requirements on the teams in terms of what work they have to do in order to participate effectively, and to meet the increasingly demanding criteria necessary to win the various medals and prizes on offer (Balmer and Bulpin, 2013).

STS scholarship has helped to identify and describe how the iGEM competition is set up to shape and perform the moral economy of synthetic biology by encouraging the sharing of material parts and associated data. The international growth of the iGEM competition has swelled the ranks of the community in terms of individual participants, countries, finances and material and informational resources. Simultaneously, the competition has also played an essential role in testing out and enforcing the values, concepts, research agendas and practices of the nascent field (Cockerton, 2011; Frow and Calvert, 2013a). The competition is regularly used by proponents of synthetic biology to represent the field in public and evidence its increasing capabilities and potential. Our colleagues were no different in this regard, as they included funding for an iGEM team in the project and actively sought to promote the team's role in the use of synthetic biology standards and so to validate that synthetic biology was happening at the university.

The choice by the academics in our project to include funding for an iGEM team as part of the project grant was intended to tie together the local work around synthetic biology with the practices of the growing international community, and also to showcase some of the research being done in the university on this global stage. Moreover, the iGEM team members were selected from engineering and biological disciplines, not only to ensure that they had the requisite skills to succeed in the competition but also to enact the interdisciplinary vision that our academic colleagues were constructing of synthetic biology at the university and demonstrate alignment with iGEM ideals.

The iGEM summer project was powerfully shaped by senior academics in the team towards the concerns of the water industry. iGEM was built into the project and shaped to meet its requirements more generally. Indeed, the team was guided towards working on the development of a biosensor, one of the key proof of principle projects that our colleagues had put into the grant application and pushed for during the industry days. As such, the team ended up working on the creation of an engineered strain of *E. coli* bacteria that could sense, using a modified version of an existing sensing mechanism, the presence of cholera bacteria in water. This would demonstrate that it was possible to use synthetic biology to create biosensors not only for cholera but potentially also for a range of other microorganisms and contaminants.

The project leader and others in the engineering departments came to describe the team's work and their apparent success in the competition as evidence of the value of a synthetic biology approach for solving water engineering problems more broadly. It became a significant part of how our colleagues demonstrated the new vision of standardisation, automation and readymades when talking with water company actors. Moreover, that the team were all novices was used as evidence of the power of these new practices. Our colleagues also used iGEM as an example of how the project had enacted the interface between academia and industry promised by the field. In these ways, the iGEM project was being used tactically as a kind of boundary object between the local concern with making synthetic biology relevant to the water industry and the broader community-forming practices of the field at large.

Relatedly, through the participation of the iGEM students in the everyday life of one of the engineering labs, we witnessed how the competition operated as a vehicle through which the materials, practices and ideologies of synthetic biology at the core were enrolled into, and linked with, locations peripherally connected to this emerging field. By virtue of iGEM, the project lab became host to a kit plate of BioBricks and a set of protocols and practices through which to do synthetic biology. The parts and protocols of the iGEM competition acted as immutable mobiles (Latour, 1987), designed to carry the norms of synthetic biology out into the world. The students and advisors thus became concerned with how to make their project fit the ethos and logics of synthetic biology and the reward structures of the competition. They had to do this from within the lab, the project, the department, the research group, the university and the UK synthetic biology community more broadly. As such, in our ethnographic collaborations with the team we observed how everyday laboratory life at the university had to be reorganised to accommodate the norms of synthetic biology in iGEM. At the same time we saw how the iGEM version of synthetic biology was modified to fit the local situation.

iGEM not only introduced a new set of materials, practices and ideals into the laboratory space, but a whole new group of bodies too. Before the beginning of the competition, this laboratory was home to a small set of postdoctoral researchers working alone on their individual bioengineering projects. These researchers were experienced bioscientists or engineers, skilled in a variety of molecular biology techniques acquired through extensive training and experience in the lab. They generally had microbiology backgrounds rather than engineering ones. With the onset of the competition, this quiet laboratory became populated by a team of six undergraduate students, a Ph.D. student and two advisors all from outside of the engineering department, all with a diverse array of skills, knowledge and experience in different disciplines and all committed to one very time-sensitive project. The iGEM team included two engineering students with scant knowledge of molecular biology and little to no experience of working in a biology lab. The remainder of the team was made up of a group of biomedical and molecular biology undergraduates with varying degrees of practical laboratory experience. For some, it was their first time in a 'real' research lab. Then there were Andy and Kate, now respectively working in and studying sociology, with their undergraduate biological training all but a distant memory. Last, the team's primary scientific advisor occasionally accompanied the team into the lab. He was a highly skilled and experienced scientist in microbiology and molecular biology techniques, and the team were hugely dependent on his guidance and advice during the project. However, this was not his lab, nor was it his department. For the summer, the scientific advisor was to be displaced from his home in molecular biology to supervise a group of students in an unfamiliar lab, in an unfamiliar discipline and as part of an engineering, rather than biological, research group. In one way or another, we all entered the laboratory as a cluster of novice bodies in unfamiliar territory.

Overcoming the body as a hazard

The students' bodies formed a key site where negotiations between how synthetic biology was enacted in the project and the broader constitution of the field played out. To explore this we examine how bio-objects, like bacteria and parts from the kit, and the bodies of the iGEM team were mutually constituted. This relationship figured their bodies as a hazard to the bio-objects as well as to other things in the laboratory and the project more generally. Indeed, such relationships, for example between team member's bodies and the bacteria they were engineering, were made explicit, questioned and resolved, as part of their own practices of problem solving, learning, interacting with advisers and working on their presentation for the competition. The relationship was acute from the early stages of the project when a series of failures became apparent, as can be seen in the following extracts from the team's lab book:

Wednesday 14th July: Transformation Failure

Neither strain grew on the Amp plates. Suggesting that either out cells were not competent or that the transformation did not work

Friday 30th July:

Transformations failed. Transformations from yesterday repeated

Monday 2nd August:

Transformations failed again.

Thursday 12th August:

Transformations failed. Possibly due to unknown resistance on the pSB1C plasmid.

Wednesday 18th August:

All transformations failed.

In trying to make sense of these failures and improve their chances of success in the future, the team and their advisors regularly discussed the possible causes of, and remedies to, their troubles. For example, the team talked about the difficulties they faced in translating written procedures and methods into material success in the lab. They were somewhat overwhelmed by the diversity of protocols available and the differing guidance and instruction offered by their various advisors. This advice ranged from how best to handle a pipette when loading a gel to judging the optimum antibiotic concentrations for bacterial transformations. Despite a wealth of codified instructions, access to existing automated laboratory equipment and standardised technologies such as PCR machines and electrophoresis equipment, the team struggled to make their synthetic biology project work. Embodied practices, it is well known, exhibit a resistance to being codified, and although synthetic biology had promised that everything would just click together like Lego bricks, for the students failure followed failure.

As time went on, however, some of the students began to develop the necessary technical and manual proficiencies to perform a range of molecular and microbiological techniques in the lab successfully. Through repetition and trial and error they slowly learned to adjust and negotiate the written procedures and anecdotal advice of their advisors in the situated, embodied context of their individual experiences of laboratory work. In this way, the students followed the counsel of one of their advisors to find a technique that 'works for you', acknowledging that practices can be multiple and that written protocols can act only as a heuristic. Students also found that some members of the team became particularly adept at performing a specific procedure. For example, one student was understood to have a 'knack for running gels.' As certain individuals began to display distinct practical and embodied competencies, a division of labour evolved within the team in which specific students would take responsibility for performing a particular technique. These divisions became quickly entrenched as the team sought to 'save time' in order to complete their project by the competition deadline in the autumn.

For the advisors to the team trained in microbiology and molecular biology, the lengthy processes of trial and error through which a novice develops the necessary embodied skills of everyday laboratory practice were understood to be an essential part of becoming a competent bioscientist:

> One of my bosses used to say you're not a molecular biologist until you've done 10,000 mini-preps and I think I've probably done 10,000

mini-preps but it kind of gives you the idea of it, it's not boring, it's the routine you have to get in and you have to sort of get green fingers....They [the students] are going to actually have to put on their lab coats and go into a lab, and it doesn't matter how clever they are if they can't use their hands and their wits, it's not going to get them anywhere...iGEM gives them that experience and I think all of them will benefit tremendously from finding out things go wrong. (Academic Molecular Biologist 2)

Failing and learning to adapt, manage and gain experience from that failure were thus seen as integral to the process of training students to become skilled biologists. In following the learning trajectory of novices in the laboratory as part of our fieldwork, we became attuned to the way in which experimental failure was made sense of by understanding the human body as a 'hazard' in the local sociomaterial ordering of scientific work. For example, early on in the summer project, one advisor demonstrated a series of basic, sterile techniques for working with bacterial cultures in the lab. Even though some members of the team already had laboratory experience from their undergraduate courses, their primary adviser wanted to start from square one. He emphasised the importance of these basic procedures to prevent the experimental apparatus from becoming contaminated by other microorganisms found in the environment. In particular, our bodies were understood to be the primary source of failure, being conduits for enzymes and microbes capable of destroying the DNA-based building blocks of their work.

The students thus learned a series of techniques to manage the hazard that their bodies posed to the successful accomplishment of their project. Managing this risk involved employing pieces of equipment such as Bunsen burners and autoclaves to create sterile workspaces and materials. It also required the students to learn particular ways of positioning and using their bodies in manipulating materials in the lab. The students were taught ways of minimising bodily contact with the apparatus when transferring cultures and reagents between vessels. They were encouraged to work quickly and smoothly by handling the equipment in specific ways. For example, they were instructed not to touch certain areas on the pipette, to angle test tubes away from the face and to use the little finger to unscrew and hold the cap of a test tube to prevent placing any of their tools onto the bench. Coughing, talking, sneezing and any unnecessary movements were all frowned upon when performing these procedures. At first, the students' actions were awkward, slow and stilted – they were clumsy bodies, a far cry from the quick, controlled techniques demonstrated by their advisors.

The incompetent, novice body was also a source of concern to the experienced researchers and technicians who shared their space in the lab while undertaking a range of other projects unrelated to ours. The iGEM team were, on occasion, accused of being the cause of problems and accidents in the lab. For the professional scientists and technicians these incidents were not simply a question of 'health and safety'. Instead, they manifested a breach of laboratory norms and represented the significant hazard that the untrained and undisciplined novice body posed not only to people's welfare, but also to the smooth running of the laboratory and the success of the experimental work being conducted therein. In this regard, the promise that synthetic biology will make it easy for anyone, let alone trainee bioscientists, to design and engineer novel microorganisms was not immediately obvious in the local laboratory. A range of pre-existing relations between knowledge, objects and people were unsettled by the team's presence. There was an emerging tension, even at this early stage in the project, between the disembodying dream of synthetic biology and the routinised and sometimes messy embodied practices more common to life in the laboratory. This tension became more visible as time passed and the team's efforts to understand and overcome their failures in the lab became more pressing. Sometimes they blamed their individual capacities and abilities, seeing themselves as bodily and cognitively lacking compared to others in the contest. At other times, it was their institution (the project, department or university) that was understood to be deficient. The team criticised the training with which they were provided, the available financial and material resources and the expertise of their advisors in the field of synthetic biology compared to other institutions in the contest. In the latter case, other teams were seen to have an advantage by virtue of access to local expertise in using standards and parts in the ways specifically required by the iGEM protocols. Whereas for our university this was their first real attempt to get a team working over the summer, others had been investing in iGEM for years.

The students' primary response to this feeling was to rely less on the labour of local expert bodies and to buy-in readymade kits from commercial suppliers. This practice was recommended by some of iGEM's core synthetic biology figures. At a European training event for iGEM teams, the major figures of Randy Rettburg and Tom Knight argued that a commercial 'off-the-shelf' infrastructure makes bioengineering work easier for the novice scientist by reducing the time spent learning and performing the embodied skills of manufacture in the lab. The companies making these products are thus important drivers in enabling students with little time and few resources to work on bioengineering projects in their summer break. Outsourcing manual lab work extends and builds upon current lab practices that often delegate resource-making activity to technicians and commercial enterprises supplying scientists with readymade materials and tools. Recall however, the ultimate ambition in synthetic biology is also to outsource the assembly of standardised parts to the automated mechanisms seen in the Gingko labs. The practice of buying in kit and delegating labour can be seen as one step toward this vision.

Thus, all iGEM students have access to readymade resources in the kit of BioBricks sent out to each team at the start of summer and the vast majority of participants outsource the further work of making new BioBricks to DNA synthesis companies. Sponsors of iGEM include gene synthesis companies, such as GeneArt and MrGene, that offer discounts to participating teams and emphasise the speed and reliability of their automated synthesising technologies. They highlight the *time* that scientists, particularly novices, can save by using their services:

> Whether you have limited cloning experience or simply want to save time, the GeneArt® Gene Synthesis service helps you move your ideas from the planning stage to the laboratory more quickly. (GeneArt, 2015)

Our team tried to adopt the practices of core synthetic biology by using these standardised processes and bio-objects. Such commercially available 'off-the-shelf' resources were understood to reduce the potential for error in their work and so could be 'trusted' and 'relied on', as the students described, in contrast to their own, hand-crafted efforts. Critically, they viewed readymade resources as crucial to 'saving time' by bypassing the lengthy trial and error processes of learning to produce functional resources and the time spent making such materials.

However, there were financial constraints on how many of these readymade resources could be purchased. Universities that can afford to support their team whether by accessing 'in-house' professional expertise and services or by providing the funds for teams to purchase resources from commercial providers are likely to have a greater chance of success in the competition (Bennett, 2010b). So although it is synthetic biology's mission to make biology easier to engineer for everyone, including neophyte scientists, a successful iGEM project is likely to have significant institutional support and expertise in this emerging field, and to be well funded with excellent corporate connections and a strong institutional legacy in the biological sciences.

In this regard, iGEM teams able to make use of institutional capital and capacities are more likely to be able to live up to the idealised image of synthetic biology. It is not just the disembodied practices of standardisation and the supposedly Lego-like qualities of BioBricks that make for a successful project. Instead, as the literature on standards more generally has demonstrated, much has to be done in the local situation to get standards to work (Fujimura, 1996; Jordan and Lynch, 1998; Lampland and Star, 2009; Timmermans and Berg, 1997). For example, in the context of synthetic biology, standardised biological parts still require adaptive tinkering and 'kludging' (O'Malley, 2009) in the lab to make things work. However, it is not only in adapting and making possible the use of these standards that institutional capacities facilitate. They also act in resistance to adaptation, and constraints upon, the successful implementation of iGEM norms.

In our project, the team's primary advisor continued to push them to acquire the embodied skills that he prized as part of becoming a competent molecular geneticist or microbiologist. His emphasis on being able to make the materials themselves was very much about becoming adaptable – being able to do good science within changing environments. This was something we found to be common in accounts from the molecular biologists in the project, and in related projects, for example the network in synthetic biology that was also based at the university. These actors talked about the importance of disciplining their bodies, emotions and attitudes in particular ways, in a similar fashion to those we heard coming from the molecular biologists. The successful control and engineering of the biological world for these academics is achieved through processes of self-transformation that can be learned only through the embodied experiences of work and failure in the lab. For them, something was lost in the process of disembodying, not just gained.

> So if everything just worked [as promised in SB], then you'd just follow your protocol and then when something happened you just wouldn't know where to start. I mean, you are like that in the beginning [of your training]. I guess its experience really and you follow a set of basics that you learn in the beginning and then you just add your experience to that as you go along and try and be a bit thick skinned when things don't work. Bugs aren't trying, you know [to upset you], don't take it personally. (Academic Molecular Biologist 3)

This disposition permeated how the advisors of the iGEM team, who were mainly molecular biologists, encouraged them to face the challenges that they encountered. It shaped how they trained them to overcome failures of various kinds. Their primary advisor wanted them to go through this process because, as he would explain, materials would not always be ready to hand. Financial constraints, another advisor pointed out, were a part of everyday laboratory life. Deliveries were delayed. Someone used your stuff without asking. You filled in the wrong form. These mundane aspects of working with bacteria thus necessitated a skilled body with a diverse range of protocols, knowledge and techniques to draw upon at a moment's notice. This was how the advisors wanted the team to overcome failure, not simply as novices learning embodied skills, but as a routine part of professional academic work.

Their vision was to create skilled bioscientists, not synthetic biologists as such. If microbiological and molecular biology practices are about adaptability, training the body to overcome unexpected hazards and constraints is essential. However, synthetic biology wishes to claim that this is unnecessary: standardised, controlled and predictable practices, its proponents argue, do not require adaptability. So participation in the manual work of laboratory life is devalued as part of becoming and being a competent synthetic biologist. To speed up bioengineering, embodied competence must be replaced by machines. This ethos is already beginning to restructure and accelerate the material-temporal regimes of bioengineering work, as we can see in bio-industrial spaces like Ginkgo and, to some degree, in the iGEM competition. However, in our project, as the iGEM team tried to enact synthetic biology in this manner, they encountered resistance of various kinds and struggled to realise fully the promise of the field to design and engineer microorganisms easily with little embodied expertise.

As we have shown, a key point of resistance in these competing norms for laboratory life concerns the relations between the body, bacteria and knowledge. The kind of hazard posed by the body is being reconfigured by emerging norms in synthetic biology. Although the body has long been a hazard to bacteria in the lab and to the effective implementation of protocols, it is now also conceptualised as a problem as part of a wider socioeconomic system. Embodied skills cost time and money that cannot be afforded. Failures encountered in the laboratory also posed a risk for the success of this particular research project, which was to function as an emblem of what synthetic biology could do for the water industry. The clumsy, novice body was an impediment to the attempts to shift the practices of molecular biotechnology and demonstrate the potential of the field as part of the promised second industrial revolution. The iGEM bodies had to be trained and managed using these long-standing practices because they were a hazard in the lab, and so this forced the advisors to mix existing ways of working with the emerging norms of synthetic biology. The body became a further barrier to be overcome in the pursuit of synthetic biology practice. If, for synthetic biologists at the core, the idea is that *anybody* can do bioengineering, once the body itself is removed from the process, for the molecular biologists who acted as advisors to the team, a successful engineer couldn't just be *any* body, it had to be a skilled, adaptable, experienced and managed body – a body that was no longer a hazard in the local situation.

Hazards to the public body

As we have argued in the book so far, our colleagues were working hard to show that synthetic biology was feasible for use in the water industry. They used various means to accomplish this, but they knew from the start that it might not be easy nor happen in the lifetime of the project. Past experiences with the industry had shown them that even robust, effective scientific research could fail to find purchase in the companies. The exuberance that they felt for synthetic biology and its bold promises was matched by their fear of failing to get it accepted. As we described in Chapter 2, the project prefaced a series of barriers to synthetic biology's success. Our colleagues thought that the public's ignorance of the value of water was one such barrier and that their naivety would stop the industry from investing in high-tech, high-risk research. They presupposed that the public would fear genetically modified organisms and that this might hinder their engineering attempts to overcome the industry's conservatism. As the Industry Broker put it:

> It's a very cautious, very conservative industry. The reason they're cautious is they have a public health remit and therefore anything they put in [to the water supply] has got to be thoroughly tried and tested, otherwise you pose a threat to public health. And in the current environmental remit everything has got to be tried and tested because not only will you pose a risk to the environment but you'll get fined and prosecuted for the contaminant, for the contamination of the environment. So they're very cautious about what they put in. (Industry Broker 1)

Our academic colleagues also anticipated that the complex of barriers to getting synthetic biology to be successful in the industry would be inter-related. As we detailed in Chapter 1, they regularly articulated a connection between the water industry's conservatism, the public's ignorance of the costs and work involved in water supply and the numerous regulatory frameworks that organise the industry's commercial and public health activities. Moreover, this was also the case for some of the industry actors, who similarly saw connections between their approach to research and development and 'public attitudes':

> I suspect that we'd want to be able to demonstrate to the customer that there is no way that [bacteria] passed through the treatment process and that the organisms are contained and do a specific job and, you know, are not released into the environment. I think if there was the potential for them to be released into the environment, then I suspect the industry would probably take a conservative approach because of the potential bad publicity associated with releasing organisms into natural systems... Seeing what the public response is to various things, including, you know, issues associated with water treatment, such as the use of chemicals, fluoridation for improving dental health, through to people's views on GM crops and GM technology in general.... that will probably give you a view of how they might view the release of synthetic biology into the environment. (Water Company R&D Manager 2)

The academic engineers and the industrial actors shared similar views regarding the connections between these various barriers. However, it was primarily our engineering colleagues in the project who wished explicitly to overcome them. In making sense of our role as social scientists, they would often position us as being the ones who might be best equipped to do this. They hoped that gaining a better understanding of 'public views' and how regulations were organised or could be changed might enable them to shift industry practices sufficiently for industrial actors to take the promise of synthetic biology seriously.

As social scientists, we ventured out into the water industry to explore how these various issues were present within the practices of water treatment. However, rather than encountering barriers like an ignorant consumer or a conservative industrialist, we encountered bodies once again, only this time, the body wasn't a hazard to the success of the engineering work being conducted. Instead, the body was at risk from the various hazards posed by water, bacteria and large scale industrial processes (Balmer and

Molyneux-Hodgson, 2013). The body of the consumer loomed large. It was a collective, public body that necessitated protection.

We first came into contact with the public body during one of our encounters with the industry, as we took a walk through a public health trail. The trail we walked forms part of the industry's public engagement and education practices. A number of the clean water and sewerage treatment facilities in the UK are public facing, in that school groups and other interested individuals can visit them to learn about how water engineering works. The public health trail told the story of how the industry has contributed to the public's health since the Victorian period. An industrial engineer guided us along the trail that began with an exhibition of various photographs and technical objects before descending into the bowels of a massive clean water treatment facility. On the wall of the exhibition space, there were pencil sketches and photos that depicted a brief history of water treatment. As we followed the engineer he pointed out one of the pictures, which showed Victorian-era children playing around a well in a busy street. He recounted the story of John Snow, a famous physician, who identified a hand-pump well as the source of a cholera outbreak in London. For the water engineer relating this story, Snow's solution was elegant in both its simplicity and its success: he took off the handle so that people could not use the well any longer. As the story tends to be told, the outbreak ended, proving Snow's hypothesis about the transmission of cholera via contaminated water, and from this event public health and water policy began in earnest. The engineer described the tale with pride. It showed why he worked there and that there was a valuable service being provided through the work of his industry.

Over its course, the trail builds into a celebration of Western engineering's conquest over bacteria, beginning with John Snow and ending with a taster of the latest treatment technologies. The dominant message from these educational texts and images is that microorganisms are dangerous and that the water industry is there to protect the body from this hazard. For example, after the romp through the history of water treatment, a series of images portray cartoon bacteria, protozoa, viruses and suchlike as ghoulish, monstrous, many-eyed anthropomorphic creatures with gloopy mouths and spikey claws (see Figure 4.1). These demonic depictions are there to play into the imagery of young minds, and, along with the story of modernity, to remind us that the industry is the hero in this narrative.

Chlorination and flocculation as industrial processes, the children learn farther down the trail, are vital to fending-off the 'bugs' that would

Figure 4.1 Demonic depictions in the water industry

otherwise harm them. These are certainly not the fragile bacteria that are so easily upset and degraded by our clumsy and hazardous bodies in the academic labs. Along the rest of the trail, the students see small versions of the apparatus used in water treatment. They learn about sand filtration and are shown how much effort goes into turning reservoir water into potable supply to their taps at home. In this regard, the ontological status of bacteria as dangerous, and the body at risk, are both connected to the water industry's emphasis on what they contribute to public health.

This story is reflected in the water companies' websites, many of which provide educational depictions of the water treatment process. They are also very much a part of the various companies' marketing practices. Companies detail the scale of the endeavour, for example, by highlighting the vast quantities of water that are sequestered from the environmental water cycle into the treatment process (Thames Water, 2015). As Yorkshire Water's website says: 'A lot goes on at our water treatment works to make your water safe for drinking' (Yorkshire Water, 2015). United Utilities (2015) has a similar message in its video for primary school children, in which an engineer standing over pipes in a reservoir says, 'Even though the water looks clean, you wouldn't want it coming straight to your taps.' Towards the end of this video, the voice-over reminds the audience that 'Chlorine is carefully added to kill any remaining bacteria making the water safe to drink and ready to send to your home.'

These educational materials have been developed because the success of the water industry as regards provision of a clean supply is understood to also be a problem for industrial actors. During our interviews and observations, the consumer of water is routinely understood to take water for granted, and to be ignorant of the value of water, only 'seeing it' when there is a fault or when water bills rise. Industrial and academic actors alike maligned the public's understanding of water as a right and bemoaned public ignorance of the expensive work that goes into supplying clean water. As a water company R&D Manager explained during an interview:

> I think they [consumers] take water for granted because, and that's a good thing because it's like a hygiene factor, until something goes wrong they think they're okay and water's always on tap. It's become part of life. ... A lot of people don't realise how complex it is to collect water, treat it, distribute it and receive waste water from houses, treat it, and discharge it back to the environment. It's very complex, and the challenge is to try and keep customers' bills low. So customers tend to worry about their bills. More and more people [in the industry] are taking an interest now, they're trying to be a bit more open and visible and arrange more communications and open days and things like that. (Water Company R&D Manager 1)

As such, companies have been investing for a while in various kinds of educational activities as part of their broader marketing strategies. They want consumers not only to have hygienic, unproblematic water but also to appreciate the consistency of the service and the cost to the industry of producing it. In this way, engagement and education have become central to the enterprise of negotiating the price of water with Ofwat, the statutory regulator, meaning that the industry opens its doors in order to convince consumers of the risk of dirty water and the value of water treatment. To them, the job of preventing hazard to the public body needs to be valued.

As we have described in Chapters 2 and 3, the industry is subject to a range of regulations that govern how it operates, and it is also financially incentivised to meet particular targets. Meeting targets on treatment failures and the number of customer complaints received, as part of the service incentive mechanism (SIM), leads to greater financial rewards for the industry and allows the companies to negotiate with the regulators to be able to charge higher prices for their services. In this way, their hazard-prevention work forms part of the regulatory cycle that shapes industrial practices of innovation.

The body enacted through the assemblage of practices of education, marketing and regulation is collective: it is the 'public body', and one that is co-produced with bacteria, which are correspondingly enacted as a mortal danger. Ultimately, industrial actors engage in political and rhetorical work to establish the value of water treatment because the practices of engineering and consumption have rendered water *too* mundane. In order to evidence the value of their services, negotiate increases in price limits and so increase profits and returns to shareholders, such actors have constructed practices that make water more visible through the co-production of bacteria, the public body and innovation. Keeping the body at risk by making bacteria a hazard is an outcome of industrial actors' practical work within a broader set of governance processes and institutional rhythms.

Hazards to industrial bodies

Our colleagues in the project felt that it was not only the public's ignorance, as they often understood it, that could pose a problem for the success of synthetic biology in the water industry. The industry broker in the project was keen to point out to our academic colleagues that R&D managers were not representative of the industry more generally. Indeed, our colleagues knew this already, since they were aware that the R&D managers, while crucial, represented a kind of 'soft case' to be convinced of synthetic biology's promise. Our engineers wanted to involve these 'science-ready' industry representatives in the project to enrol them into the promissory narratives of synthetic biology as a route into the industry more widely – to be gatekeepers. As one of our engineering colleagues explained:

> To some extent we're going to be working with more of the free minded people in the water industry. The water industry is very diverse, [although] in general it is quite a conservative industry, there are some people, particularly in the R&D area who are more open to ideas. Starting with them first we're trying to slowly work our way down the R&D departments and then into the companies themselves. Even if it is just raising awareness. What people in the industry will be interested in is the potential of synthetic biology in the industry, and then they might think about the problems. But to encourage them to be interested you have to give them an awareness of the potential before they disregard something to start with. (Academic Water Engineer 2)

The engineers also wanted to get the R&D managers on board so that they could help with one of the project objectives, to generate future grant applications. Industry would do this for example, by writing letters of support. At the same time, and in the longer term, they realised that in order for synthetic biology to be successful in the industry they would have to convince accountants, managers and other executives, as well as the many industrial process engineers who actually ran and maintained the infrastructure. However, our broker was certain that the process engineers would be just as ignorant of synthetic biology as the public were. At one meeting he described process engineers as button pushers, with no expertise, and explained that they would probably be fearful of synthetic biology.

Our ethnographic work in the water industry facilities meant that we encountered process engineers of various kinds, and whilst they were not experts in genetic engineering or in water research and development, they nonetheless exhibited plenty of embodied expertise that was crucial to the successful operation of the infrastructure. Moreover, their practices, situated within water industry facilities, manifestly shaped the co-production of their bodies and the bacteria with which they worked.

This is most clearly evident when considering the scale of the infrastructure. After moving through the education trail we described earlier, we walked out into the clean water treatment works proper. Whereas the exhibition space was quiet and gently lit, the treatment works was different altogether and the shift in atmosphere was palpable. There were large pipes everywhere, some as thick as ancient oak trunks and others more like metal branches, that ran along the floor, across the walls and up onto the ceiling. It was cold, because everything is made of concrete and metal and largely filled with cold water. And the deep hum of that running water murmurs through the air, erupting into a deafening roar as the pipes come to an end and water cascades down concrete waterfalls, sloshing and foaming into gigantic vats. There is no escaping the connotations of dirtiness when you look into the tanks. The shitty brown substance floating on the surface coats the large brushes that scrape up the floc and slop it off the edge, as clean water moves on in the process exiting via the bottom of the tanks.

At the same time, the facility has a number of chemical smells that change as you move through the staged process of filtration. These come from the various compounds that are added into the water during its treatment, which variously create the floc, change the pH, remove contaminants and so forth. At the end of the process, chlorine is added to the water to help to stop bacteria from growing in it. Various practices in which process engineers are engaged in the clean water facility are thus primarily organised around the removal of harmful bacteria and other contaminants. In this regard, their everyday encounters with microorganisms are framed by these practices, so that it is entirely sensible to view them as hazards.

The treatment facilities are places in which dirtiness and cleanliness, danger and health sit side-by-side, but in which a vast number of practices and their associated chemicals, tools, machines and so on work to maintain the borders and ensure that dirty water becomes clean and clean water does not become dirty. The binary in the water infrastructure between sewerage and potable water is embodied within the material arrangements of such treatment works. As Douglas (1966: 4) argues, the boundaries between these parts of the world, between polluted and pure, are under pressure: 'ideas about separating, purifying, demarcating, and punishing transgressions have as their main function to impose system on an inherently untidy experience.' The processes through which the water engineers maintain these distinctions are routinised in the organisation of the space, and in their bodily encounters with these spaces, tools, and so forth. Danger and risk are under constant control. At the same time, these practices themselves can introduce a certain amount of danger for those responsible for their performance. Some of the chemicals are dangerous, the heights at which the process engineers have to work can be frightening and the depth of the tanks and their frothing brown liquids all pose particular risks to the engineers' bodies.

This is even more the case in the sewerage treatment works. Whilst the clean water facility was largely dry and clean, save for the tanks of floc, the sewerage treatment works was a mess. In this everyday site, we followed process engineers treating sewage and managing the facility. Here the pipes, waterfalls, tanks and vats were all present, as with the clean water works, but they were largely outside, exposed to the environment, and so the floor was covered in mud. There was mud absolutely everywhere, and it cakes everything it touches. Hoses are kept near to entrances to buildings for washing off the mud as your clothes, shoes and gloves are periodically caked, cleaned and caked again. When we approached the site for the first time it was clear that we were relatively unprepared. Our industry guide took us to the boot of his car, where he revealed two pairs of boots, and a number of plastic bags which he uses as part of a process for keeping his normal shoes clean and his wet boots separate from other items. Our guide had clearly routinised his encounters with mud into his practices of work.

The atmosphere at the sewerage treatment works was dense with a vapour that wets you like tropical air and seems to fill your nose, mouth and lungs. Much as in the clean water facility there were gigantic tanks of water, but here they all contained huge, dense swirling masses of bacteria that are used to treat sewage as it moves through the processes of filtration before being released back into the environment. The tanks are deadly deep and have oxygen bubbling from the bottom to aerate the water and keep the indispensable bacteria alive. This, the process engineers noted as we crossed the pools, makes them even more lethal if you should fall in: 'You'd sink quicker than we could get to you.' An untrained body in the geography of the sewerage works is awkward. It feels clumsy just as it does in the lab, but rather than this clumsiness being a hazard to bacteria, in the water facilities a clumsy body is itself at risk. Bacteria are the hazard and the body is fragile.

Moreover, the bacteria are a health risk. One of the engineers in the sewerage works explained how process engineers in the works have to get ill before they become accustomed to the place:

> When you first come to site you get ill being around this atmosphere. But your immune system takes to it after a bit. When I've been to other sites, I've gotten ill before, from stomach bugs. (Water company process engineer 1)

As Process Engineer 1 described it, the human immune system 'takes to it' or as another said, 'Once you've been ill, you build an immunity' (Process Engineer 3). Lots of the anecdotes we were told in these field sites were organised around the way in which novice engineers or visitors were ignorant of the potential to be infected by the bacteria. One such story, for example, concerned a group of electrical engineers who had been contracted to fix a problem with an aeration tank and who had all refused to wear masks over their mouths and then became ill, one-by-one, day-by-day until none of them was left to finish the job.

In this regard, the process engineers' bodies are at risk from bacteria by virtue of how they are enacted within the practices of sewerage treatment. Their embodied understandings of bacteria as risky to their bodies were not an outcome of ignorance or of a distrust of science, but of their situated relationship with bacteria, which were a different thing in the water facilities than they were in the laboratories.

Conclusion

Synthetic biology's promise to overcome the faltering industrialisation of past biotechnological projects is bound up with efforts to shift relations between the body, knowledge and bio-materials such that the human body is increasingly edited out of scientific and industrial situations. However, in the academic context that we examined and certainly within iGEM, which continues to have a prominent role in displaying the success of synthetic biology, the body remains salient. Whilst the untrained body poses a threat to bio-objects, experiments and ultimately the success of research projects, it is the development of embodied competencies that then enables the scientist to manipulate and work with the contingencies and unpredictability of biology and laboratory life successfully.

This is not to say that synthetic biology is not wearing away the existing practices that enact the relationship between the body and microorganisms. Certainly, at the core, it would seem that there are emerging technological practices, using robotics and so forth, that appear to reduce the role of embodied knowledge in the laboratory. However, there is some evidence from our project and encounters with core spaces, for example through iGEM and its jamboree, that suggests that how current practices are represented in public is strategic. The robotics displayed on the Ginkgo website and the emphasis on a kit of parts travelling around the world in iGEM help to present an image of science done without bodies. However, the bodies remain in everyday practice, if not in its representation. At present, we would argue, synthetic biology, both at the core and in the periphery, is a partial implementation of certain engineering principles, mixed up with and reconfigured alongside more long-standing molecular biology techniques and their associated embodied skills.

We have shown that, despite emerging changes in the material-temporal patterning of laboratory work, the body remained central to how synthetic biology was enacted in the periphery. Specifically, we have explored how existing practices configured different kinds of bodies as 'hazards' or 'at hazard' in relation to particular places, projects and bio-objects.

Moreover, whilst positioning the body as the cause of failures, the future vision of disembodied synthetic biology comes with a range of risks and its own modes of failure. The embodied knowledge created in the laboratory, as genetic engineering is currently practised, is about more than getting things done. It is about learning a particular kind of self-management, which allows for a reflexive awareness of the agency of things, allowing scientists and engineers to cope with failures to control microorganisms and so to be resilient, prepared and capable of adapting to contingent and changing circumstances. In the same way that knowledge gets 'black boxed' and locked in to scientific and innovation practices, as is well documented in STS scholarship, there are also ways in which knowledge gets 'locked out' through changes to embodied practices, with equally unpredictable and complex results (Marris *et al.*, 2014). How synthetic biologists will learn to cope with themselves and the practices and materials with which they work when things go wrong is currently unclear.

Even if the relation of the body and bacteria is changed in the practices of synthetic biology laboratories, it will not be enough to shift how this relationship is enacted elsewhere. The body remains salient in innovation systems and consumer spaces. In the water industry, for example, educational, marketing, regulatory and professional practices have come together to constitute risky relations between the body and bacteria. Ties such as these might form a kind of ontological barrier to the introduction of synthetic biology outside of the lab, and so might have significant implications for the ability of our colleagues and others working to bring about the synthetic biology vision at the academia-industry nexus.

Synthetic biology materials, tools, practices and visions are negotiated in relation to multiple embodied norms of understanding and managing risk and failure in everyday academic and industrial life and in how these are entangled. As we have demonstrated in the previous chapters, the field is being performed through the reconfiguration of existing practices, whether those are regulatory, industrial, academic or scientific. Here we have shown that existing relations between the body, knowledge and bacteria continue to resist and rework how synthetic biology is being practised and adopted.

Coda 4
Reciprocal Reflexivity

Andy Balmer: What do you think the role of social science is in an interdisciplinary project such as this?

Academic Environmental Engineer 1: Well having spoken to Susie just sort of generally, in coffee shops or whatever, I think understanding how science is done would be useful. It's not something I've ever stopped and thought about. It's just something that you do. But if you understand how you do it, maybe there are things that you can learn to help to do it better. So I think that's more of an internal reflection on what we do.

Concerns with collaboration, co-production and reflexivity animate our specific methodological strategy to studying and engaging with synthetic biology. As discussed earlier, this collaborative orientation aligns with a broader post-ELSI approach to embedding reflexive practices into everyday scientific and technical work. In this context, efforts to work *together* with scientists and engineers are motivated by the goal of 'opening up' (Stirling, 2005) the politics of scientific knowledge production with a view to reshaping technological practices. Responding to these sensibilities in our own work, we sought to develop methodologies that encouraged co-productive modes of engagement and 'reciprocal reflexivity' (Balmer and Bulpin, 2013; Balmer *et al.*, 2015; Calvert and Martin, 2009) with our scientific and engineering colleagues on the project. Our interactions with the iGEM team became one of the most successful and fruitful efforts in collaboration.

Our participation in iGEM was shaped by our academic interest in changing practices, everyday scientific knowledge production and community-building mechanisms, as well as our particular institutional positioning on the project. In the latter respect, all three of us were tasked alongside our scientific and engineering colleagues with ensuring that the iGEM team was successful, albeit in different ways. Success was measured in multiple, but interrelated metrics by academics on the project, for example, in terms of winning awards in the competition, satisfactorily evidencing the 'doability' (Fujimura, 1987) of synthetic biology, demonstrating its applicability to water engineering problems or just learning something useful for their careers and lives more broadly.

Our involvement in iGEM manifested in Andy and Susie becoming 'advisors' to the undergraduate students and Kate enrolling as a member of the team. In the wider context of social scientists' 'upstream' enrolment in synthetic biology, and Susie and Andy's respective institutional positioning as co-investigator and postdoc on the project, in many ways we started out as full participants and members of the world we wished to study. Furthermore, not only did we seek to nurture co-productive, reflexive relations with the scientists and engineers on the project but we also actively pursued strategies that were intended to shift, or at least to reframe the emerging epistemic practices. This impacted significantly on how we engaged with the iGEM team.

Although, as discussed in Codas 2 and 3, our sociological interests and ambitions in synthetic biology did not always easily map onto our colleagues' expectations for social scientists' involvement, the field at least was open to our participation from the beginning. One of the ways in which our social scientific input into iGEM was valued by the academics on the project was, unsurprisingly perhaps, in relation to the 'human practices' element of the competition. 'Human practices' (or 'policy and practices' as it has now become known) is one part of the iGEM criteria for winning a gold medal and has its own specific prize within the contest. Put simply, human practices asks students to consider the ethical and social dimensions of their technologies with an increasing emphasis placed on demonstrating how these considerations have shaped their scientific projects.

One of the key objectives of our approach to supporting our iGEM students with human practices was to develop a collaborative, co-productive methodology that embedded reflexivity into the everyday life of the students over the iGEM summer project. We talked to the students about the kinds of questions we were interested in as STS researchers involved in synthetic biology. We taught them a little about STS theories. Emerging out of this discussion we worked with the team to develop a 'human practices' methodology designed to track how their own practices and roles in the team took shape and shifted over the course of the project. Through collaborating on the production of an experimental visual device designed to map out the different social and technical factors shaping the project, we engaged in some highly productive reflexive discussions (Balmer and Bulpin, 2013). We explored how and why particular decisions had been made; how these decisions had affected the project; the social, material, technical and institutional constraints the team had encountered along the way and how their individual roles had shifted and consolidated over time.

Alongside these human practices activities, Andy and Kate also became intimately involved in the everyday labours of the lab. We added our hands to the daily grind of manual work, whether pipetting solutions or sourcing equipment and materials for the next stage of the team's experiments. We accompanied the team to meetings with their advisors, articulating the kind of difficulties we were experiencing in the lab. We worked with the students to put together presentations of their project for their advisory team in the university, for their competitors (and collaborators as iGEM prefers to position them) at UK get-togethers and ultimately to an international audience and judging panel at the MIT jamboree. We spent time with the team outside of work socialising during our lunch breaks or over a beer at the end of a long day. So, by the time we arrived at MIT in Boston in that November, donning our branded team hoodies and frantically putting the last-minute touches to the presentation, we had become full members of the club and heavily invested in the team's success.

In this regard, we saw ourselves as true collaborators on the iGEM project. We were making synthetic biology happen, even if this had not exactly been our immediate aim. We had, to some limited extent, helped the team to make a biosensor for cholera. Our collaborative relations had contributed to co-producing knowledge and materials *for* the project and *about* the project. We had also succeeded in introducing greater reflexivity into the everyday life of the iGEM team. For example, we witnessed how our efforts to 'open up' the situated and contingent nature of the team's practices influenced how some of them began to talk about, and make sense of, their individual and collective experiences of participating in the competition. A particularly acute example of this reflexive shift was manifest in how the team discussed their everyday difficulties and failures in the lab and ultimately made these public at the MIT jamboree.

As described in the preceding chapter, failure was an all-too-common experience over the summer. As time passed and the team continued to struggle in overcoming their difficulties, the affective tenor of the iGEM experience shifted from one of initial excitement and anticipation to one of frustration and disappointment. Although the team eventually did get enough evidence to suggest that they had succeeded in making a biosensor, they encountered huge struggles on the way and ultimately were left unsure whether they had really managed to do 'synthetic biology' or not. At first, the students blamed themselves and turned inwards to make sense of their difficulties by attributing them to personal failings in their academic skills and scientific abilities. This individualised culpability that students feel for the success or failure of their iGEM projects is tied to an underlying meritocratic logic underpinning the competition and academia more broadly. In the context of a field that promises that the use of standardised biological parts will make bioengineering work easier, the failure to build, easily, a new biological 'machine' is even more keenly configured as an individual failing. Add to this that the groups of students competing tend to be at the very top of their game in their educational trajectories, and the emotional impact of failure in the competition can take a heavy, often personal, toll.

However, through our interactions and discussions with the iGEM students, we began to situate their difficulties in the project in a broader context. We helped them to explore how problems in the lab were connected to institutional constraints. As the promise of a gold or even silver medal began to fade from view, the students prepared themselves for managing their disappointment at the jamboree by reflecting on the institutional inequalities between the teams. We accompanied them in watching presentations made by some of the 'powerhouse' teams in the competition and discussed the differences that were apparent between their experiences of the contest and our own. Our iGEM team also expressed certain scepticism over the presentations they saw at the jamboree. In particular, they questioned to what degree the students had 'help' in their work and to what extent they were involved in deciding what kind of application their team would pursue. These critical reflexive practices were engendered by preceding discussions in which we had explored some of the various factors that had shaped the iGEM team's 'decision' to work on a biosensor.

An intrinsic part of our methodological approach was therefore about managing and making sense of the affective dimensions of changing practices in the context of synthetic biology. In particular, engaging in these kinds of collaborative, reflexive practices placed the scientific 'self' within the dynamics of shifting epistemic practices. Our methods of unpacking the social, material, political, technical and institutional entanglements constituting the iGEM team's project situated the scientific self in a network of relations extending out beyond the lab. We helped the team to take 'care' of themselves in relation to the complex sociotechnical networks in which they were working, as part of their production of knowledge about microorganisms, and as part of moving ahead in their own careers and lives.

Attending to care of the self and affective practices has received very little attention in STS (Kerr and Garforth, 2015), but it is an integral part of connecting changes to local, everyday practice at the 'periphery' to the community-building strategies used at the 'core' of new technoscientific fields as well to more global shifts in economic and political discourses, governance and innovation writ large. The co-production of scientific practices, scientific communities and scientific subjects is an explicit element of educationally oriented programmes such as iGEM. Our academic colleagues recognised the intrinsic connections between the successful accomplishment of the project, moving towards synthetic biology practices and transformations of the self that are backgrounded in the competition:

> the iGEM jamboree makes them think that in order to be successful it has to work, which I think is unfair because there's so much more that's gone into being a successful iGEM team...I do think there's a whole educational experience that goes into iGEM. (Academic Environmental Engineer 1)

The educational experience of the iGEM students was not only about learning embodied skills or new knowledge about biological processes and practices but was also about learning to manage the emotions involved in doing scientific work. Although scientists will readily admit to the frustrations and difficulties as well as triumphs and joys of their work, the explicit management of the emotional labour involved is primarily understood as an individual responsibility, and one that should be erased from descriptions of how knowledge gets made, at least in formal accounts. Moreover, an implicit expectation about our involvement as social scientists in technical projects is often that we will deal with the affective elements of making such projects 'work', for example, in facilitating collaboration between disciplines and managing disagreements when they arise (Balmer *et al.*, 2015).

Certainly in the context of iGEM some of our most significant roles in the field were of a pastoral nature. Often we took on the role of 'counsellors', considered sufficiently independent from the project to allow students (and their advisors) to air their frustrations and grievances about their work. At other times, we were more concerned friends, sharing and discussing our experiences of recent events. This mode of engagement with the students and academics brought us into the heart of the scientific work and constituted us as colleagues rather than mere contributors. Helping to care for making knowledge was a productive but also challenging and sometimes problematic experience.

As we navigated increasingly close and collegiate roles with the iGEM team over the summer we began to reflect on how our emerging personal and professional commitments to the team were also making it difficult to retain a more critical stance towards the project and to synthetic biology. As the jamboree drew nearer, we became increasingly wrapped up with helping the team to finish their project and prepare for their upcoming presentations. Our days became filled with long periods in the lab with the students frantically trying to assemble the biosensor and collect evidence that it worked. Our nights became dedicated to helping the team in constructing reports for the wiki and proof reading draft presentations. At the same time, our sociological questions and interests began to disappear from view until it seemed we had succumbed to the 'seductive power of technoscience' (Nordmann and Schwarz, 2010) or, more traditionally, had 'gone native.' We had become unable to distance ourselves critically from the enactment of synthetic biology and had begun busying ourselves with making iGEM happen according to the norms of the competition because we wanted the team to succeed in the ways that they wanted and that the advisers wanted. Care led us into carelessness, to some degree.

Nonetheless, nurturing collegial relations with the students had been instrumental in generating the trust and openness that was so important in developing the kind of co-productive interactions that we celebrated earlier. We considered our work with the iGEM students to be a positive, *successful* example of a post-ELSI mode of collaborative engagement in contrast to others' more arduous and ultimately abortive experiences (Rabinow and Bennett, 2012). The discomfort that we felt in retrospect, given that our role had almost become 'novice synthetic biologists', was largely due to not knowing exactly what kinds of responsibility we should then bear for the success or failure of the team and of synthetic biology more generally. Of course, this was just one team and one year of the competition, but our enrolment in enacting synthetic biology was a little troubling. At the very least it sensitised us to the importance of striving to care without becoming careless about those things that lead one to caring relations in the first place.

There were also ways in which our presence in the iGEM team took on a more disruptive role. In striving to retain a critical stance towards iGEM and synthetic biology earlier in the project, we left open an ambiguity in the collaborative relations that we were developing with the students (and their advisors). Our critical distance sometimes put these relations at risk as the students were sometimes unclear as to what our own research was saying about them. Were we moles there to report on the failure of the team, or on the failure of synthetic biology? Would we be divulging some of the everyday secrets that get papered over in formal presentations? Over time we did provide answers to these suspicions, as we shared the notes we were taking, began to write up papers, gave presentations and so forth.

In a similar way to our disruptive encounters with some of the academics on the project described in Coda 3, we also found that our presence on the iGEM team sometimes upset, disquieted and frustrated the students. We felt that our constant questioning became tedious to them and sometimes our physical presence in the laboratory felt unwelcome. Certainly, our novice bodies were also significant hazards within this space. Although Andy and Kate had training in molecular biology techniques, they were not nearly as skilled as were our undergraduate colleagues. There was also an issue with the amount of time that our human practices work took up on the project. There was sometimes a sense that too much time was being diverted from the more important matters at hand in the lab or in the modelling work. Despite the team's energy, commitment and involvement in the human practices work, it remained difficult for them to articulate the value of this work when weighed against the time it had cost them. Indeed, towards the end, as we approached the jamboree, the time we had to spend on human practices seemed to the team to have been at the cost of spending more time building and characterising their biological part and perhaps had come at the expense of a silver medal.

They did, however, see the value of the work during the competition, as judges praised them for the excellent reports they had written and for the inventive approach they had taken to human practices. After the competition was over, we occasionally caught up with some of the students and it had become a little clearer to them how they could use the knowledge we had made. One of the team found it helpful during her PhD interview to be able to discuss scientific practices in a more reflexive fashion. Another found his frustrations in the laboratory during his PhD more comprehensible, given his ability to reflect on their location within complex networks of practices. There was, in the long term something to be said for time spent on reflexivity, but the ways in which such work becomes valuable remain outside the usual metrics for scientific excellence.

That being said, there were also problems with encouraging the students to take a more reflexive, critical stance to the competition and the field. Although there were certainly benefits to this work in helping the team manage the emotional costs of their participation, if we were not responsible for puncturing their initial enthusiasm we were at least culpable in cultivating a more cynical attitude amongst the team. By illuminating the ways in which many aspects of the project's development were out of their control, we ran the risk of promoting the notion that there was no possibility for change or agency in their work. We debated whether we might actually constitute a hazard to their chances of success in the contest when we resisted or critiqued the normative frameworks structuring the competition. Perhaps if we had engaged in a more familiar ELSI analysis and focused on the technical objects the team produced, we would have stood a better chance of winning the human practices prize.

In this regard, thinking about care and affect in the midst of attempts to produce reciprocal reflexivity, we must consider our contribution and responsibilities towards shaping the practices and experiences of our collaborative partners. With what kinds of cost do new post-ELSI modes of engagement come for us and also for our colleagues? What we might constitute as a successful collaborative endeavour might be constituted very differently by our science and engineering colleagues, who are likely to be working with different metrics of success. How post-ELSI collaborations make sense of success and failure will be important factors to consider in the future design, enactment and review of such experiments.

5
Enacting Ontologies, Failure and Time

> To read some accounts of synthetic biology, the ability to manipulate life seems restricted only by the imagination. (Kwok, 2010)

Introduction

In Chapter 1 we made a case for understanding synthetic biology, as we put it, 'in situ'. By this we meant that what the field is, how people try to do synthetic biology and with what consequences are importantly shaped by the situation in which it is enacted. We highlighted some sensitising themes on every day practices, promises and ontology, arguing that by understanding synthetic biology at the level of everyday practices we might be better equipped to moderate the bold promises made by the field's core proponents, and to see more clearly how ontologies are reconfigured practically as people try to bring about change.

We sliced the project three ways, examining barriers, bacteria and bodies to illuminate some of these issues around the emerging technoscience. We now take another three slices, this time through the book, in order to carve out a description of how the engineers in the project tried to change what they were doing in order to make synthetic biology happen. We explore the first slice, 'ontology', looking at how objects, like bacteria and barriers, were enacted. We also argue that synthetic biology itself had to be enacted within changing practices. Ultimately, our colleagues were not entirely successful in enacting synthetic biology, at least not as they had observed it in their encounters with the core. Accordingly, we also then slice the project by 'failure', examining how this was managed and with what consequences for the affective dimensions of everyday life in the project. We tie failure in to the mechanisms of synthetic biology more broadly in order to open up a critical relationship between the core and periphery. We argue that how the core and periphery are entangled raises ethical issues that have yet to be articulated as such. Finally, we take a slice through 'time', probing the ways in which timing and tempo played out differently in different practices, with important implications for why the realisation of the big promises made by proponents of synthetic biology might not be as timely as anticipated. Through these considerations of our situated study we further contribute to the developing accounts of synthetic biology in social sciences, and particularly in STS. We also treat synthetic biology as a case through which to explore some important contemporary STS themes. In each section we outline the existing literature in STS and then weave this through our case study, developing both the case and the literature as we do so.

Enacting ontologies: sticking things together

There is a major trajectory in STS regarding objects and materiality, which has recently developed into a concern with how different ontologies of objects or things are connected to situated practices. As we pointed out in the introduction, STS has become more or less comfortable with mess (Law and Singleton, 2005) and multiplicity (Mol, 2002) as regards the description of objects in different situations. The developing approach to ontology in this literature places an emphasis on how practices enact the world differently. De Laet and Mol (2000), for example, provide an account of the multiple identities of the Zimbabwe Bush Pump, in order to argue that across different situations the precise arrangement of relations constituting the pump shift around. The pump is flexible. It is adapted and it adapts to the uses of different communities, as communities change and as new elements are incorporated through repair. So what the bush pump is, exactly, depends upon how it is co-produced through local practices of enactment with various other actors. In another context, Law and Lien (2012) describe the different fish farming and scientific practices that they encountered when tracing the connections between 'slippery' Atlantic salmon and various human actors. They take these to be evidence of different salmon, performed through different practices.

In placing practices and performativity right at the heart of these accounts, STS scholarship also serves to sensitise us to the ways in which descriptions are made only for now, and that further, continuous work, by a range of actors, has to be done to make things more durable.

Woolgar and Lezaun (2013: 325) explain that 'the ways in which differently enacted entities come to seem to be the same 'thing' is the upshot of active practical work rather than a reflection of any innate commonality or characteristic.' It is in how different ontologies are 'choreographed' (Cussins, 1996) or 'coordinated' (Mol, 2012) that commonality is produced, and the emphasis so far has indeed been on how similarity or sameness in objects is created across different enactments. Ensuring that the specificity of the labour involved in maintaining sameness across different enactments is recognised, the co-production of practices and things has to be conceptualised differently depending upon the situation. For example, De Laet and Mol (2000) argue that the bush pump should be understood as a 'fluid object' by virtue of how it gradually changes and is changed as part of different practices. In this way, 'an object or a class of objects may be understood as a set of relations that gradually shifts and adapts itself rather than one that holds itself rigid' (Law and Singleton, 2005: 339). Similarly, when describing the messy and erratic object of 'alcoholic liver disease' in various practices of diagnosis, treatment and care, Law and Singleton propose that it should be understood as a 'fire object.' Whilst the fluid bush pump changes slowly through incremental adaptation and repair, the fiery disease leaps around less predictably, and with more dire consequences.

Despite such descriptions of different kinds of objects, there are notes of caution regarding the relationship that this multiple and messy approach should have to the traditional, philosophical or canonical project of ontology. If the implication of making a 'turn' to ontology was understood to be an instruction that STS should look to the abstract, then Woolgar and Lezaun (2015: 464) are at pains to point out that 'Our attention should always be directed to the webs of practices that enact a particular reality. To the extent that abstract talk of ontology detracts from this task, a "turn to ontology" would represent an unwelcome inflationary move.'

Lynch (2013) also argues that we should be less concerned with developing an ontology writ large, abstracted from the empirical sites in which STS research is conducted, and that we should instead focus on how questions about the status of objects, their sameness and difference, are worked up in a given situation and with what implications. This more ethnomethodological angle he terms 'ontography', placing it alongside 'epistemography' (Dear, 2001) and 'ethigraphics' (Lynch, 2001) to denote the way in which such matters are simultaneously ontological, epistemological and ethical. STS scholars, he argues, should examine how actors of various kinds make use of existing methods for dealing with such issues within their everyday practices. He asks, 'Can a particular category, estimate, or measure of what is in the world be separated from the practical, conceptual, and political means through which it is implemented?', thus adding that ontological matters will also always be practical, rhetorical and political for the actors entangled with them.

This developing account of ontologies as situated within practices is helpful to our own explorations of how actors in our project tried to change their practices to fit better the emerging norms and promises of synthetic biology. Ontologies were central to their efforts. Synthetic biology has, as we have described throughout the book, developed around the use of engineering epistemology to control and predict biological phenomena more easily. In this regard, synthetic biology is explicitly about enacting bio-objects like bacteria in new ways by changing epistemic practices.

Our own approach to these kinds of issues has much in common with Lynch's advocacy of 'ontography'. We examined synthetic biology through mundane, everyday scientific, engineering, academic and industrial practices. We charted how these linked to different ontologies of barriers, bacteria and bodies. We drew attention to the political and rhetorical dimensions of this work, as regards both governance and promissory discourses and more local institutional politics and individuals' strategic descriptions. In this section we borrow from the emerging research on ontologies in STS in order to bring out these issues across all three empirical chapters of the book. We also contribute some novel terms and findings to this literature.

Some scholars working to produce accounts of the multiple ontologies of things are keen to stress that in order to attend to the specifics of our cases we should multiply up our terms and adapt them to our local situation (Mol, 2012). We agree, although we find the existing use of the term 'choreography' in Cussins' (1996) research and subsequent scholarship to be instructive. One of the key contributions that choreography can make is to help us understand the temporality of ontologies, as regards their ephemerality and what implications this might have for changing practices. We will explore this term further in later sections of this chapter on 'failure' and 'time'.

However, we also need to maintain our interest in how things become more stable and in the kinds of work that are done to accomplish this. Attending to ephemerality in the enactment of things does not mean that we must assume that everything is popping in and out of existence from moment to moment. There is some consistency and continuity in the world by virtue of the material, practical, rhetorical and political webs spun through everyday practices. We propose that the term 'sticky', found in Molyneux-Hodgson and Meyer (2009) can be useful here. We extend the use of this term through its associations, such as stickiness, attachment, stuck and so forth, and add to these the term 'cluster' to describe how some things get stuck together in quite messy fashions.

Collectively, these various synonyms for stickiness are helpful in understanding the ways in which actors seek to organise ontologies in a more fixed fashion, by stabilising practices and norms, with the implication that ontologies can potentially become more solid. On this note, we contribute to the strong lineage in STS descriptions of accomplished consistency. The term 'sticky' and its relatives, however, also gesture to how this happens to varying degrees of success, and take account of the more recent emphasis on mess and difference. Not everything sticks equally. Things can be sticky like sticky notes or like cement. In other words, some ontologies can be stickier than others. Some stick together against your will, like getting two fingers stuck with superglue or getting the 'hook' of a song stuck in your mind, whereas others get unstuck, fall apart and become detached in the process. Drawing on these terms, we now develop the theme of ontology as it relates to changing practices in our colleagues' experiments with synthetic biology. We explore the enactment of things that are material and conceptual and, in the later section on time, we also treat 'synthetic biology' itself as a thing to be enacted.

As regards material things, in Chapters 3 and 4 we examined various practices in which bacteria and bodies played a role. For example, we evidenced how their ontological status as hazards or at hazard came about through their locally situated practical enactment. The entanglement of bodies and bacteria was very much about how practices were organised in different situations, perhaps becoming most evident through our comparison of the laboratory and the water treatment works. Bodies were hazards to bacteria in the lab, whereas the roles were reversed in the treatment facilities. In the clean water works, for example, bacteria took the form of floc, which was dirty and gooey because of the introduction of flocculation coagulants. This meant that the water tanks were hazardous. The same was true in the sewerage works, this time because the bacteria were enacted as massive communities of swirling and bubbling brown sludge as they interacted with the tank, the water, dissolved oxygen, organic nutrients in sewage and other microorganisms. It was not only human practices that produced these particular relations between bodies and bacteria. The materiality of the world also intra-acts (Barad, 2007; Mol, 2012), making bacteria different in different practices not only because of how human actors change what they do but also because they themselves are entangled in shifting relations with other material objects. This meant that each space we encountered required different practices for managing the entanglement of bacteria and bodies in order to control potential hazards in both directions.

Extending this analysis we also explored the way in which the concept of the public body was configured through wider sets of practices regarding the privatisation of the water industry, its profit-making mechanisms and the regulation of the sector. Through these practices the public body became vulnerable. Water services were set up around this ontology and sustained it. The public body at risk was very sticky because of its significance in the water industry's promotional, marketing and public education practices, all of which were important in the profit-making and value structures of the water companies. Practices in water services also entangled this enactment of the public body with the enactment of bacteria, which became monstrous, dangerous and in need of control.

These relations between bodies and bacteria shifted across different situations as sets of practices from governance, industry and academia were differently mixed. This caused some trouble for our colleagues' aims in making synthetic biology stick. The ontological enactment of bacteria and the body in synthetic biology is significantly different to that found in these industrial sites. The emerging practices in the field certainly offer control over bacteria and render them valuable. They tie labs and business together by promising improved prediction and speedier innovations. This was appealing to our colleagues and formed part of how they tried to stick together this ontology with those in water engineering. However, the promise of putting genetically engineered bacteria into pipes and water sources, and possibly also into customers' supply, posed a significant challenge. For the industry, bacteria are not meant to be in water. The water companies' sole purpose is effectively to clean water of microorganisms and other contaminants. As such, our academic colleagues struggled to get the new ontology of bacteria in synthetic biology to replace the ontologies of the industry, due to how they were clustered through practices, and this struggle, in turn, proved to be an important limitation to bringing about any change in broader practices of innovation.

Our colleagues were also involved in trying to stick together particular ontologies of other conceptual objects. For example, in Chapter 2 we showed how regulatory, water industry and academic practices enacted the water companies as 'conservative' and the public as 'ignorant' consumers. This meant that the industry and the public were enacted as 'barriers' to innovation in our project. Barriers became a collective term; it stuck together particular ontologies of conceptual and material objects into clusters by virtue of the entanglement of practices of regulation, innovation and knowledge production. The ontological cluster of barriers is common to academia, industry and governance. It helps actors at the interfaces to conceptualise problems and work together, for example by determining particular constraints or limits on action. So like boundary objects (Star and Griesemer, 1989), such ontological clusters help to manage difference by providing a common language, a 'shared ground' (Traweek, 1988) of norms, and a quick way to conceptualise problems across different spaces. But unlike the 'interpretive flexibility' that has become central to STS accounts of boundary objects (Star, 2010), ontological clusters also stick things together in messy, inflexible and less desirable ways.

For example, some ontologies can be difficult to detach from a cluster, as we found once we began trying more forcefully to alter the ontology of the public. We wanted to shift how they were enacted as 'ignorant', and so also as 'barriers', in order to open up the possibility that publics could be better engaged in the sociotechnical politics of synthetic biology and also of water innovation. We showed that this was difficult because of how this enactment of the public was stuck to other ontologies, like 'conservativism', 'over-regulation' and so on, through regulatory, corporate and scientific practices, so that efforts to change one thing meant also trying to unstick a number of others. In this way, trying to pry off or rework one ontology risked others becoming detached or indeed our colleagues becoming unstuck from their industrial partners. Trying to alter how things were stuck together in the cluster meant potentially losing the ability to speak the same language, or challenging shared norms, and sticking with the industry was an absolute priority for the engineers. At industry days, therefore, they found themselves working to synchronise promises, affects and practices, but having to operate with the same old ontological cluster of barriers in order to remain tied into the industry and its governance. Overall, then, there were ways in which things did not stick or stuck too well. Ontologies of (ignorant) publics, the (conservative) industry, (over-)regulation, (dangerous) bacteria and (vulnerable) bodies persisted. Ontologies from synthetic biology would not stick and, in some regards, became quite 'slippery' (Law and Lien, 2012).

In this regard, consistencies across synthetic biology and the water industry as regards the use of engineering epistemology and practices, rhetoric, politics and promises did not necessarily create consistencies in the enactment of various ontologies. In part, this was because the organisation, scale, functions, intra-actions and rhythms of the material world were also entangled in these situations. Whilst they showed some continuity through enactments in engineering practices across the academia-industry nexus, their differing material intra-actions also contributed to the co-production of things and practices, meaning that engineering was not uniform across all situations. So whilst there are some epistemological and material consistencies, the worlds of engineering are still multiple and messy. In light of this, we can see that synthetic biology's objects and concepts had to be enacted within this existing mess of practices, and in a way, our colleagues were stuck with what they had.

The implication for synthetic biology more broadly is that the adoption of engineering ontology and epistemology in order better to control microorganisms in laboratories and manufacturing processes provides no guarantee that the field and its objects will neatly slot into industrial engineering practices and markets. There is still much opportunity for failure in this emerging technoscience. We now carry over our concerns with ontology into two further sections, the first dealing with failures and the second with time. In both of these slices through the book, we are interested in how synthetic biology was enacted in the periphery, through its relations to the core, and with what implications.

Enacting failure: the onto-affective dimensions of changing practices

Within scientific communities, the topic of failure is conspicuous by its absence. For example, Pinch and Collins (1984: 522) describe how scientific literature edits out many aspects of everyday laboratory life:

> Details of the experimenter's health, the date of the experiments, the motives, interests and emotions of the experimenters, are also absent. The failures, preliminary runs, aggravations, breakdowns, financial difficulties, family and time pressures are not reported.

Instead, published accounts of scientific work align with a 'canonical model of science which does not allow for a failed experiment' (Pinch and Collins, 1984: 537). Yet, as early ethnographers of scientific culture have attested, failure is part and parcel of experimental work and scientists develop specific mechanisms for understanding and managing failure in their everyday professional lives (Knorr Cetina, 1999; Collins 1992 [1985]).

Sociological debates over the place of failure within the sciences and how scholars should approach this aspect of scientific life were front and centre in the early days of STS. The formation of the 'Strong Programme', as part of the emerging 'sociology of scientific knowledge' (SSK) was set up in resistance to how past institutional sociologies of science had dealt with distinctions between 'true' and 'false' scientific theories. Failed scientific paradigms were designated as 'false' and attributed to social and cultural interests or influences that did not align with the normative framework of the scientific community. On the other hand, accepted scientific knowledge was understood to relate to an objective 'truth' in the world, which could be accessed through adherence to rational, scientific methodologies. Theories and paradigms that failed to gain traction within the scientific community became excluded from the boundaries of science and labelled as 'non-scientific' (Gieryn, 1999). The Strong Programme proposed that sociologists should resist such scientific realism and instead follow new methodological guidelines in which both 'truth and falsity, rationality and irrationality, success or failure,' should be approached in the same way (Bloor, 1976: 7). This 'symmetrical' method situated failures *and* successes as outcomes of the complex, social fabric making up scientific knowledge production.

Although the Strong Programme has been criticised from multiple angles, not least by members of the STS community for a social determinism that belies the influence of the material world (Latour, 1992), an impartial, 'symmetrical' analysis of successes and failures in scientific and technical work has remained an important touchstone in the field. From social constructivist accounts of technology (Bijker *et al.*, 2012) to actor-network theory (Latour, 1987), scholars have approached the study of successful or failed technological artefacts or scientific knowledge as a result of the practices that enact it as such.

Despite the preponderance of failure in these early sociological descriptions of everyday scientific work, everyday struggles and failings have been only lightly woven into the ethnographic study of laboratory life, the creation of new fields of knowledge production, the governance of knowledge and the enactment of ontologies. The role of failure in organising these relations at the level of everyday practices has not yet been fully addressed. However, elsewhere in STS scholarship, there are more developed accounts of failure and its relation to bringing about change in scientific practices and governance.

As we outlined briefly in the introduction, and have made use of throughout the book, the literature on the 'sociology of expectations' and 'contested futures' (Brown *et al.*, 2000) has demonstrated how promises about technical innovation are used to configure current relations in anticipation of such imagined futures. In the context of pharmacogenetics (Hedgecoe and Martin, 2003), for example, promises were used to enrol actors around particular visions of the future from industry, patient groups, regulators and so forth. As Hedgecoe and Martin (2003: 355) put it:

> Visions provide a framework within which the future shape and application of a technology are constructed, as they act as both an aid for decision-making and a focus for the mobilization of actors and resources. In this way, new technologies, new industries and new ethical, legal and social problems are coconstructed and mutually shaped.

This is particularly the case with new and emerging technical fields, where quite radical promises about the future are used to create and consolidate new networks of actors and so to engage in new kinds of practices (Brown and Michael, 2003). Expectations are used in 'defining roles and in building mutually binding obligations and agendas. At the most general level we can understand expectations to be central in brokering relationships between different actors and groups' (Borup *et al.*, 2006). Much of this work relates to the kinds of arguments being made in the literature on enactment of ontologies. STS accounts of promises generally argue that they form a part of the generative relations out of which novel technosciences and objects are created.

Setting promises against past failures in order to distinguish current work from that of the past is a common strategy in the use of expectations to shape the present. This was very much central, for example, in the context of promises about the potential of xenotransplantation (Brown and Michael, 2003). Although scientists working in this field were aware that previous scientific promises in gene therapy had not been realised, some were still exuberant about xenotransplantation, and continued to make big promises about the future, despite their familiarity with this, and a host of other failed futures. Borup *et al.* (2006: 290) explain this by reference to a kind of amnesia, common amongst actors working towards the development of novel technosciences:

> expectations of technology are also seen to foster a kind of historical amnesia – hype is about the future and the new – rarely about the past – so the disjunctive aspects of technological change are often emphasized and continuities with the past are erased from promissory memory.

Brown and Michael (2003), borrowing from Collins (1988), anticipate a difference in how close such actors are to the 'core set' and how familiar they will be with the range of uncertainties inherent within emerging practices. However, the continued adoption of promises in new techno-sciences, set against a multitude of failures to live up to previous ones, seems difficult to explain if those closest to the development of a new science are those most likely to understand its limitations. To make sense of this, they propose that we understand actors working at the core as having to assume two different roles: scientist and entrepreneur. Switching between these two roles allows actors constructing a new field to negotiate competing pressures to represent their work and practices in particular ways, in order successfully, or not, to enact those very prac-tices and enrol actors in the periphery into the field. Promises and past failures are entangled in how novel technosciences emerge.

In focusing our empirical lens on the everyday life in a synthetic biology project, we encountered and experienced promises and failure in multiple guises. We witnessed how promises from the core were made use of in the periphery, and of how failure was papered over or made sense of. For example, we saw how promises were used to broker rela-tionships between academia and industry and to secure funding from government and the RCUK. In this regard, promises and failures were not only enacted through but also ordered and connected the micro-level practices of material life in the lab, the meso-level of academia-industry relations and the macro-scale of policy interventions. In fact, despite its promises, everywhere we turned failure loomed large in synthetic biology.

Most notably, following precedent, synthetic biologists at the core, and so also at the periphery, position the field and construct its emer-gence as a remedy to a complex of current and past failures – a dense entanglement of technological, material, disciplinary and bodily faults and deficiencies. In particular, as we have described throughout the book, proponents of synthetic biology promise to succeed where previous molecular biology and genetic engineering projects have failed. These bioengineering disciplines are understood to have failed in real-ising their promise of 'translating' research into industrial applications. As we described in Chapters 2 and 4, for synthetic biologists, the cause of this failure is identified as a kind of moral and practical failing within the biological community, making molecular biology a discipline at the mercy of nature's vagaries and the body's limitations and so unable to harness and take control of the technological and economic potential of biological matter.

Relatedly, as we have relayed throughout the book, synthetic biologists also strategically connect the disappointment of past biotechnological promises to ongoing political concerns regarding another kind of failure – that of the economy. They are able to do so through recent political and academic manoeuvres that we have outlined regarding the situation of science and technology as central drivers of the knowledge economy. Further, in the current economic atmosphere of austerity, in which research councils are ever on the brink of funding cuts, science and engineering are increasingly organised through their potential to contribute to jobs and growth. Synthetic biology has been explicitly connected to this particular vision of science in the UK Synthetic Biology Roadmap (SBRCG, 2012). In this vein, the field has also been positioned as part of a global competition, in which a fear of failure to keep up with the international 'bioeconomy' and maintain national competiveness animates both political decision-making and synthetic biology practices.

Whilst the risk of failure in synthetic biology is primarily about not realising the industrial future it promises, in the water industry failure is far more immediate and has more serious implications. The fear of failure in water industry spaces is acutely connected to risks to public health regarding the potentially catastrophic consequences of failing in their mandate to protect the public body from dangerous microorganisms entering the water infrastructure. Such potential industrial failures are also connected, through regulatory regimes at the national and international level, to global challenges, for example, in how climate change, storms and water shortages are linked to the material failure of pipes, allowing microorganisms into the water system and so also to the possibility of financial sanctions if companies fail to meet water quality standards. Our project was thus constructed around such sociotechnical challenges and failures of the water infrastructure and, as we have described, this impacted significantly on how synthetic biology manifested in the local situation.

One of the key drivers of the synthetic biology project at the university was a past record of failed attempts to translate scientific research into water industry spaces. To some degree, the project took shape around a fear of failure stemming from previous bruised encounters between academia and industry. Failed pasts became constructive of the fear of failed futures, through which researchers anticipated problems and barriers to synthetic biology's success. The spectre of failure manifested in concerns over public ignorance, fears of genetically modified organisms (GMOs) and the conservative attitude of the water industry regarding high-technology innovation. A preoccupation with the kinds of barriers that might impede collaborations between academia and industry was motivated by a simple question – how likely was the project to fail?

Of course, in what ways failure was defined and how the consequences of failure were understood differed between participants in the project. For some, failure was primarily conceptualised in terms of experimental success, for example, whether they were able or not to assemble the biosensor in the lab successfully or to create models that would accurately predict water flow. Experimental failures, in turn, posed a risk to the success of future careers, for example, in terms of inhibiting the 'cycle of capital' (Latour and Woolgar, 1986) of publications and funding upon which scientific careers depend. Laboratory failure risked undermining the potential of what synthetic biology could do for the water industry and so threatened both academics' and industrialists' trust in the promise of the field. Everyday failures became increasingly perilous as they endangered future collaborations between these two spheres.

This represents an entanglement not only of anticipated futures in which promises are realised but also of anticipated failed futures, as well as the common rehearsal of the failed promises of the past. In this way, the promissory visions of synthetic biology, for example around grand challenges and economic growth, are partially shaped by the practical fallout of the failed promises of scientific futures now past but also *anticipated*. Although the sociology of expectations literature stresses the performative character of promised scientific futures in enrolling interest and stimulating investment, and so illuminates the disillusionment and disappointment that follow scientific 'hype', it has had less to say on how such affective features of the failures of past promises shape present everyday practices and fears for a failed future. In our study, we found past failures, failed promises and feared future failures were not only mobilised as part of the discursive equipment of the emerging field but were also a key factor in shaping and shifting everyday practices of doing technical work and changing affective relations.

This can be seen most acutely in how everyday failures are erased from epistemic work through the practices in which synthetic biology is being constructed. Whilst failure cannot be avoided, and is indeed a constituent part of well-established disciplines, there is a great pressure in emerging technological fields to demonstrate early successes and hide the struggles through which any advances are made. Where novel, experimental work ventures into unknown territories of knowledge production to develop new skills, practices and objects, the risks taken are great and the failures frequent. However, lots of effort goes into concealing and managing the constitutive role of failure in the pursuit of new knowledge and the creation of an academic community.

Training is a good case in point. Science students learn about the models and metrics for success and failure in their disciplinary training and what kinds of affective displays are considered appropriate in each case (Traweek, 1988). Scientific practices shaping the reporting of success and failure also shape the scientist's experience of these phenomena. Training teaches students to remove the contingencies and failures from their formal accounts of their work (Delamont and Atkinson, 2001) and to experience the vagaries of nature as a personal failure to control the experimental assemblage (Collins, 1992 [1985]: 167). As such, the personal and emotional life of science is deeply bound up with how scientific failure is enacted. The training of students in the iGEM competition exemplifies this and evidences some of the ways in which affective practices are entangled in trying to enact new ontologies.

In the case of iGEM, the failures and struggles of the students are hidden in the often slick team presentations given to the judges at the annual jamboree, although there are some notable exceptions. Despite informal chats between competitors and reports on team wikis recounting the trials and tribulations of doing an iGEM project, these stories of failure disappear by the time the students reach the jamboree. It is perhaps unsurprising that failures are largely hidden in a competition that awards prizes and medals based on criteria designed to align with the objectives and visions of the maturing and highly skilled academic synthetic biology community.

Through the reward structure of the competition format, the proponents of synthetic biology make very clear what yardstick is used to measure success and failure in iGEM. Highlighting any difficulties or failures in team's projects would go against the grain of the promissory visions of synthetic biology more broadly, and undermine the role of the competition in demonstrating the potential of the field. To talk about material failures in the lab is to threaten the engineering promise of synthetic biology to replace the 'trial and error' methodology of traditional bioengineering approaches with readymades. Moreover, the introduction of standardised parts and engineering practices is promised to make bioengineering work easier precisely for novices like iGEM students. Thus, there is the implicit expectation within the competition that projects should look 'easy'. Failing and struggling in the lab is part of the bioengineering past and not the synthetic biology future.

At the jamboree, winning presentations tell alluring tales of ambitious ideas translated into material success in the laboratory with the promise of innovative applications just on the horizon. These kinds of stories echo the bold claims and linear narratives told in research grants and scientific policy. Over time, teams learn from past participants and advisors at their institutions that iGEM rewards those teams that comport themselves in a manner that reflects the future image of the field. It is noteworthy that at the beginning of the competition in 2006, there was a prize for the 'best conquest of adversity' in recognition of the difficulties of 'getting anything at all to work' (Carlson, 2010: 89). However, this disappeared the year following and has not been a part of the competition since. In making this change diversity and failure were rejected as an officially recognised part of doing synthetic biology in iGEM, although they continue to be a significant part of many students' experiences of participating.

Demands on the iGEM teams have also increased as the synthetic biology community has sought to tighten and stabilise the normative framework of the competition to align with the field's evolving objectives and vision. Notably, the increasing amount of characterisation of standardised parts that is now demanded of iGEM teams in order to do well in the competition reflects the ongoing failure to realise the functional standardisation of biological components in the lab, not just in iGEM but in synthetic biology more generally. Over time, the work that the iGEM teams do in submitting new parts and improving existing parts contributes to the continuing improvement of the BioBrick™ kit for the next year's competitors, and also serves to reify the synthetic biology community through the story of iGEM's success.

Successful projects incorporating extensive characterisation work and significant material contributions to the registry are mostly accomplished by a select few host institutions that have invested in the competition year on year. At universities that have committed to iGEM, and synthetic biology, over the last decade, teams and their advisers have accrued significant capital and capacity in the field. This is reflected in their students' continual success in the contest. Institutional memories play a key role in facilitating this through the involvement of students and staff with a history of participation, including team members who have gone onto become advisors and even judges at the jamboree. In these universities, the sociomaterial practices of synthetic biology have become routinised into the organisation of their laboratories and workspaces. In many cases, these institutions have built significant resources, reputations and connections in synthetic biology. They can support their iGEM teams through, for example, providing sponsorship, access to commercial infrastructures and industrial partners, as well as the knowledge and expertise needed to anticipate and meet the growing set of standards and expectations set by the competition. In this way, a select group of institutions have come to dominate the iGEM contest in winning the big prizes year on year and their projects have become emblematic of what a successful synthetic biology project looks like.

The import of all of this is that each year, as the competition attracts a new collection of teams from around the world with different institutional histories of participation in the contest, projects are judged according to the community's increasingly demanding criteria and in relation to emblematic accomplishments. This creates a cycle in which the norms and materials of the competition circulate across the world, are adjusted and circulate again, year by year. Although the judges are aware of the differences in cultural capital and resources between the teams and make efforts to address this in their appraisals, the institutional inequalities between the teams remain, and they, along with the various failures of the teams, are effectively erased.

It is thus becoming harder and harder for those teams without such institutional capacities to succeed in the competition. As each year's projects are calibrated against other's successes, accounts of failure and local constraints continue to be edited out. The narratives of success encouraged and rewarded in the competition conform to, and perform the vision of, synthetic biology as making bioengineering 'easier for anyone, regardless of training, to construct novel biological systems' (Billings and Endy, 2008). As Swidler (2001: 96) says of changing cultural practices more generally:

> the establishment of new social practices appears not so much to require the time or repetition that habits require, but rather the visible, public enactment of new patterns so that 'everyone can see' that everyone else has seen that things have changed.

In this manner, iGEM functions as a public enactment of synthetic biology's promised futures in which novices can create new products using standardised parts in a short period of time. It is a spectacle and attracts an increasingly international audience of students and professional academics, as well as industrialists and governance actors.

However, this cyclical process conceals the tremendous amount of work and failure that has gone on 'behind the scenes' at multiple institutions and the huge amounts of capital required in order to shift existing sets of practices, sociomaterial spaces and institutional relations to cope and continue when things fail, and so to try to meet the increasingly demanding community standards.

Synthetic biology not only constitutes a moral economy in connection to the open source exchange of parts and use of standardisation practices (Balmer and Bulpin, 2013; Frow and Calvert, 2013a), therefore, but also requires participants to perform their work and experiences of iGEM in ways that align with specific visions of the community. This means that a promissory economy is interwoven with the moral economy, which together circulate a specific vision of synthetic biology by rewarding or erasing particular articulations of the past and future.

This moral, promissory economy is also accompanied by an affective economy, which values and rewards certain feelings over others and so encourages and discourages different kinds of affective and emotional displays depending on to what extent they fit with promissory visions of the field. Failure, shame, embarrassment, frustration, anger, disappointment, abjection, hopelessness and apathy all play significant roles in the doing of an iGEM project, particularly in the periphery where resources, institutional legacy and adviser expertise are more constrained. However, these are not the kinds of emotional and affective displays that correspond with the bold promises of success and visions of novice bioengineering fuelling the competition and the field more broadly. Students learn to hide not only their failures but also this compendium of more troubling and unpleasant feelings in their formal interactions with each other, the judges and the synthetic biology glitterati that attend the competition most years.

This applies not only to iGEM but also, perhaps to a lesser degree, to the professional academic community. As we detailed in Coda 3 and Chapters 3 and 4, our engineering colleagues also had to negotiate what constituted synthetic biology by evaluating and adapting their own definitions, practices and use of promises in relation to those coming from the core. For example, as we described in Chapter 3, attendance at conferences and workshops in the core were central to our colleagues' reconfiguration of their own work, professional trajectories and efforts to change how their department was positioned within institutional power relations. The SBX.0 conference became particularly vital. Going to SBX.0 every few years meant that our engineers had encounters with the shifting norms and standards for practices at the core, and so also with the ever increasing demands for what constituted synthetic biology proper. In 2011 at the SB5.0 conference that we attended with our colleagues, many core actors from the USA and UK were in attendance.

Some of the most prominent core actors made speeches in plenaries, on panels and as part of their role as chairs and conference organisers in which they drew attention to exemplar objects, projects, academics, labs and companies represented at the conference and in the field more broadly. As founder and President of the BioBricks Foundation, Drew Endy wrote in his welcoming message:

> [Since 2008] we have seen significant scientific and technical advances, including full genome synthesis, reliable synchronization of multi-cellular genetic oscillators, and opiate precursor biosynthesis. We have also experienced increased politicization of the field, including the U.S. Presidential Bioethics Commission's consideration of synthetic biology, and ongoing popularization of such work through activities such as the iGEM. (Endy, 2011: 9)

As with many academic conferences, the selection of speakers was limited by its scale. Whilst the majority of participants were represented in the programme only by a poster, of which there were over 300, others were singled out as having something particularly important to report. These selections represented attempts to formalise what synthetic biology was, at that time, in academia or industry and established a hierarchy amongst participants as to who counted as a true synthetic biologist and who did not. These speakers almost uniformly came from massive, elite institutions having huge pots of funding from industry, government and other sources.

For example, George Church, from Harvard University and an extremely prominent geneticist and engineer, began his talk, jokingly, with a 'conflict of interest' slide, on which were plastered over 50 logos from government, educational, charitable and industrial institutions that funded his research. His subsequent slides highlighted particular projects, companies and universities that had contributed to the grand challenges in synthetic biology that he saw as being 'well under way.' As he said, 'we're making progress, and these are truly the grand challenges, solving energy and infection' (Church, 2011). His talk was peppered not only, as would be expected, with findings from the rather extraordinary research that his lab was conducting but also with pronouncements on where practices, knowledge and things were at in the field more broadly, and where they would be in the future. In the question and answer session, for example, he predicted that the cost of synthesising DNA would fall in a similar fashion to Moore's law in microelectronics. Moore, co-founder of Intel, had stated that the number of integrated circuits on a transistor had doubled each year since the 1970s. Church predicted, as others in the field have also done, that the costs of DNA base synthesis would fall just as rapidly, so that at some point in the future it would cost only billionths of a cent per base pair. There was much discussion about these kinds of topics throughout the conference, of where technology was heading and of how the economics underlying the field's vision for success might develop.

There was talk of the core actors' failures and struggles in their presentations, but these were largely about limitations in the field, and not in their local situations. They were about where things were at now in science, not just in their labs, and so indicated that things would improve over time, for everyone. The conference acted as a platform for core actors to report their research, showcase their funding successes, diagnose the challenges, predict the future and get people excited. These activities are common in academic conferences, but here they were also ways in which 'synthetic biology' itself was being standardised, in order to constrain not only what could count as good work in the field but what counted as being part of the field at all.

This was very exciting for our engineers, since they loved the technical details and were animated by what synthetic biology could do. It was a way of getting further integrated into the community. Indeed, some of their close colleagues in the UK spoke at the conference and chaired sessions. However, it was also frustrating and disappointing. It was clear that we did not have the same kinds of resources – financial, human, material and institutional – to conduct work at this scale and this quickly. Whilst they acknowledged these limitations to us, they were less eager to talk about their own struggles during chat around their posters or over conference meals and breaks. They did not have quite as much pressure as did the iGEM team to conceal their struggles, but there was certainly a degree of performance, of maintaining an image that synthetic biology was happening in their labs and that it was working.

In both iGEM and in academic professional synthetic biology, bringing about changes in practices and in the ontological status of biological objects, from DNA to microorganisms, involves knowledge-making practices that are also affective. The concealment of institutional inequalities, struggles and failures exacerbates and heightens the emotional burden of the already demanding experimental work whereby participants often internalise their difficulties as individual inadequacies. As such, the maintenance of the moral, promissory economy of iGEM and synthetic biology more generally comes at a significant cost, and one with an important, but largely ignored, affective dimension, particularly for those at the periphery.

The work that goes into changing existing practices of knowledge production and the ontologies of biological objects in established epistemic workspaces thus requires significant emotional as well as financial, political and material investment. Efforts to reconfigure extant sociotechnical relations around new sets of promissory visions, practices and ontologies are fraught with difficulty and failure. As people try to adapt and change their practices to fit the imaginaries and norms of synthetic biology, whether as part of the iGEM competition or a professional research project, they bear the risk of both the project and the field failing to find traction. They must keep their hopes up, keep working on their own promises for their own futures and manage their feelings and fears of future failures. In this regard, actors in synthetic biology are constructing part of their own emotional and affective lives in relation to the enactment of the field's ontologies through changing practices, and so also are constructing the field through the affective practices of their everyday lives. These are the onto-affective practices central to bringing about change in a novel sociotechnical field. They are a key to understanding how changes in sociotechnical practices, imaginaries and ontologies are enacted through the practices of everyday academic life.

Enacting time: (a)synchrony and change

We have begun to examine time in synthetic biology in the previous sections, documenting some of the temporal dimensions of practice, how promises and futures are enacted or not and the importance of annual events like iGEM in the distribution, arrangement and experience of failure as actors work to bring about change. In this section, we explore time more explicitly, bringing out some more of the temporal dimensions of the project and of synthetic biology more generally.

Sally Wyatt (2007) suggests that STS has paid less attention to time than other areas of social theory, and wonders in what ways an analysis of 'situated time practices' (Wyatt, 2007: 824) might illuminate our accounts of sociotechnical systems. Ethnographic approaches are inherently situated in *places* and so it is quite common for ethnographies of labs, problems and so forth to situate observations accordingly. However, only a few texts bring the temporal dimensions of ethnography explicitly to the fore. Garforth and Cervinková (2009) have helped to explore how time might be brought into how we situate knowledge production. They examine the character of multiple different regimes of time in academic research and how 'heterogeneous forms of practice, discourse and ordering constitute different *timescapes* for researchers' (2009: 169). They document some of how scientific practices play out under different circumstances, of power relations, of disciplinary focus and of career stage.

Their analysis follows Adam's (1994) emphasis on two aspects of time, namely 'timing' and 'tempo'. As regards timing they examine how academics 'fit everything together, coordinate actions and synchronise agendas' (Garforth and Cervinková, 2009: 169). When it comes to tempo, they are concerned with 'questions of the speed, pace and intensity of practices and institutional change' (Garforth and Cervinková, 2009: 170). They also highlight the acceleration of science, regarding the ways in which nations are more explicitly oriented now to trying to bring about the anticipated benefits of the knowledge economy, and how researchers on the ground must try to stay on top of deadlines and manage increasing workloads. In this way, the study of the everyday dimensions of academic life and of how people try to make changes in their practices can benefit from considerations of temporality in the form of both timing and tempo. Moreover, 'within the timeframe of the everyday, flexibility and autonomy must also be understood in the context of the increasing importance of audit and performativity in academic life' (Garforth and Cervinková, 2009: 212). It is vital to make sense of how time operates at different levels so that we can see how everyday practices, affects, pressures and opportunities are entangled with broader discourses and larger sets of institutional practices and relations of power in the choreography of temporal regimes.

On this note, Hannigan (2006) also points to Adam's work to highlight the centrality of time when thinking of emerging uncertainties. He sees a congruence between Beck's risk society thesis (Beck, 1992) and how we try to negotiate uncertainty in sociotechnical futures. He argues that we cannot rely on current understandings to deal with emerging technologies, given the absence of precedents and the array of ambiguities involved in the present. In such circumstances, rhetoric from powerful actors becomes the mode of making sense of the novel and of the future, and can carry much influence over policy-making processes (Hannigan, 2006: 149). It will be important to understand how futures are constituted in relation to power, particularly in the context of governance, when considering the emerging and uncertain field of synthetic biology. Drawing on the terms we have deployed in the book, we could rephrase Hannigan accordingly: core actors help to constitute temporal regimes of governance through strategic use of power relations in order to enact synthetic biology in particular ways.

Indeed, in only a very short time synthetic biology moved from being a speculative musing (Roco and Bainbridge, 2002; Swierstra *et al.*, 2009) to a 'great technology' in which governments were heavily investing (Willetts, 2013). The field seems to epitomise the trope of the 'sheer scale of scientific change' (Franklin, 2008: 11) in the contemporary. This relates directly to the epistemology of synthetic biology, as we have examined throughout the book. At the core of the enterprise is an efficiency project. Changes in epistemology and practice are intended to speed up the engineering of microorganisms, which is crucial to the economic promises of the field and its ability to ameliorate global challenges before their effects are irrevocable. This predictive biology, · by taking the guesswork out, compresses time toward the delivery of applications. The industrialised and manufacturing slant of synthetic biology's rhetoric, promises and emerging practices thus recalls the traditional production line, and so also a Taylorist vision of productivity. The instrumentalist, economic, global framing of synthetic biology requires that practitioners are always busy designing and making stuff, doing so more efficiently, saving money, saving the planet and rescuing the economy.

Nationally, governance time regimes are also shaping synthetic biology's timing. By virtue of the alignment of certain visions of the future in economic governance and the field's epistemology and rhetoric, synthetic biology is being hurried through. The UK Synthetic Biology Roadmap comes with its own diagrammatic and innovation timeline, concentrating efforts and investment, to bring about the field in a timely manner. The changes promised by the field are scheduled through these timelines and have to be brought about 'on time' in order to warrant the investments made in the field and to secure further funding. This is also connected to how synthetic biology's sociotechnical imaginary is constructed, so that the field is understood to play a role in the UK's global economic competitiveness. The fear of 'falling behind' international competitors, primarily the USA and China, is a common refrain in governance and academic circles, and certainly helped to shape how the RCUK has invested in the field.

To begin with, funding was relatively limited, and only for short-term projects. Our own project fell into this category, as did the initial networks in synthetic biology (Molyneux-Hodgson and Meyer, 2009). However, before long, huge amounts of money were invested in a short timeframe to create six synthetic biology research centres, each lasting for five years, as well as the related networks in industrial biotechnology and bioenergy (NIBBs), again lasting five years. The Roadmap and other governance devices were important in bringing about this escalation and hurrying-up of the field's enactment. The sociotechnical imaginary (Felt, 2015; Jasanoff and Kim, 2013; Pickersgill, 2011) of synthetic biology is one in which the epistemology of the field and the nation's economic future are co-produced. Audit culture and governance time regimes are therefore performative, they order sociotechnical change according to neoliberal mechanisms of competition and the contrivance of markets, making synthetic biology in particular ways, which in turn reifies those practices.

How different time regimes were fitted together or did not synchronise in practice were important features of how these broader rhythms were negotiated in the quotidian experience of our project. As such, the timings and tempo of synthetic biology and the water industry are important to consider in how practices were being changed, or not. As we explored in Chapter 2, our engineering colleagues hoped to overcome barriers to innovation in the water industry, using synthetic biology as a 'hard case'. They put together synthetic biology in particular ways, experimenting with certain materials and practices and using certain parts of the terminology and rhetoric, in order to fit them together with the ways in which innovation operates in the water market.

Recent changes in the governance of these markets had meant that water companies were now being asked to plan for long-term research and development. The same kinds of global challenges appearing in synthetic biology were also being incorporated into how the water industry was regulated. The field was pitched to water company actors as a way in which they could begin to think long-term about these problems and how they might invest in high-tech science and engineering. However, in order to synchronise the time regimes in this fashion, it meant pushing synthetic biology's promises further into the future. What the field could deliver for the water industry became a long-term goal, since everybody knew that intractable issues like climate change and pipe leakage were not going to be solved any time soon.

How global challenges like climate change and so forth manifested materially within governance and industry practices also made a crucial contribution to this asynchrony. Indeed, the massive scale of the infrastructure alone meant it that would take a long time to implement any novel solution, and cost a fortune. Local practices in water treatment facilities and the expertise required in these situations would take time to reorganise, particularly if high-tech solutions were going to be implemented. As we explored in Chapters 2 and 4, a significant amount of time would be required for process engineers to begin using synthetic biology products in a treatment works or when testing water for pathogens in rivers, reservoirs, pipes and so forth. The set-up of the water companies as natural monopolies also meant that there was little incentive to spend profits on improving services, particularly since customers were already quite satisfied with water as it stood. In fact, it was part of the challenge for the project that water was largely invisible to consumers because of how it was enacted through these practices. As we have argued, synthetic biology actually represented a potential diffi-culty since the introduction of microorganisms into water posed a chal-lenge to how the water companies fulfil and market their commitments to public health. Making water more visible could backfire. So, without a consumer imperative, the drivers for change in the water industry's R&D practices came primarily from the pressure to respond to global challenges. Such challenges have an immediacy, but by their scale and complexity require long-term thinking and change.

However, the regulation of profit structures in the new service incen-tive mechanism did not much address such issues. The contrivance of competition thus failed to tie innovation practices to global challenges adequately. This misalignment of concerns and practices in regulatory and industrial spheres shaped how synthetic biology's promises and sociotechnical imaginary was integrated into the local situation. The temporal regime embedded within the practices and rhetoric at the core of the field was reconfigured, in the local situation, so that the episte-mological promise of driving innovations from the lab to industry at an ever increasing rate was stretched out over a longer time period, but with the consequence that nothing had to change immediately, and indeed the water industry actors barely lifted a finger towards bringing synthetic biology closer to market. Whilst promises and sometimes affects could be made to fit together, things did not quite sync up as expected or hoped.

This example demonstrates that promissory narratives are not received and adopted without change in local situations. Instead, they are worked on and altered in order to make them relevant and synchro-nise them with other promissory narratives. The term 'choreography' (Cussins, 1996) becomes useful here, particularly when mixed in with current work in STS on the enactment of ontologies through situated practices, and with all of the rhetorical, political and practical work that we described in the first section of this chapter. Choreography is helpful because it provides a distinctive temporal sense to ontological enactment and, of course, promissory discourse and sociotechnical visions are about the entanglement of the past, present and future. We can extend the sense of the term, for example by highlighting how work has to be done to organise the movement of things in coordination with each other. It means temporality and rhythm. There is the possibility that things might fall out of sync; there can be slip-ups, errors in timing and alignment. Choreography can thereby fit quite neatly with some of the current work in STS that tries to highlight the ephemerality and fluidity of ontologies as enacted through practices. As we have encountered in Chapter 2 and have just described in this section, our colleagues certainly had to engage in practical rhetorical work, of talk, giving presentations, demonstrating cases and so forth, in order to choreograph the visions of the future and needs of the present in synthetic biology at both the core and periphery with the visions of the future and needs of the present in the water industry. Global challenges became crucial to this work.

In exploring this kind of choreography across different practices, we have to understand that the performative nature of promissory narratives and sociotechnical imaginaries perform sociotechnical fields like synthetic biology in situation-specific ways. What synthetic biology was in our local situation was dependent upon our colleagues' choreography of practices and ontologies, like those of the public or industry, and of barriers, bacteria and bodies, at the confluence of synthetic biology and the water industry. In this regard, whilst promissory narratives like those found in synthetic biology might exhibit sameness within emerging sets of practices, as actors use promises to help to perform such sociotechnical fields at the core and the periphery, their transmission is also importantly a part of how difference is brought about. As diverse visions of the future are brought together through the situated mixing of practices and attempts to change ontologies, they might also produce asynchrony in the present. This impacts on how well synthetic biology travels from the core to the periphery. In other words, how futures are choreographed through situated practices can reconfigure the situated performance of sociotechnical fields in the present.

It is clear, then, that different timings, or temporal regimes, exist in different areas of governance, for example in the governance of synthetic biology there is lots of haste, but in the governance of the water industry time is stretching out. This means that temporal regimes in governance and industry may not synchronise quite so easily with the promised epistemic reordering of laboratory practices. Even if robots and standards help engineers to design, construct and even manufacture bio-objects more rapidly, there is no guarantee that this will translate into faster uptake of academic research into commercial markets. As we saw in our project, academics have taken it upon themselves to shift how they engineer microorganisms, but they have far less power to reorganise regulatory and industrial practices, which are organised around other concerns and values that predate the emerging field.

However, it was not only to accomplish industrial goals that our colleagues were working to change their practices and make synthetic biology happen at the university – it was also part of their everyday academic life. Here we can perhaps see more clearly how working at enacting ontologies becomes co-productive in a different regard, by virtue of how it is tied up with the lives and identities of scientists. When De Laet and Mol (2000) term their bush pump a 'fluid object,' they also suggest that 'its story tells us that actors, technologies as well as the engineers involved with them, may be fluid' (235). This relates to their description of the engineer who helped to create the bush pump and how he himself tries to remain fluid by rejecting his definition as the 'inventor' or the 'owner' of the pump.

This is quite an unusual observation in that STS has regularly shone the spotlight on 'things', but has taken relatively little interest in academic and professional identities. In contrast, engineering studies has devoted significant attention to the latter, sometimes at the expense of the things with which engineers work. However, Downey and Lucena (2004: 400) do argue that engineering identities have to be understood in relation to material entities, since 'The identity politics of engineers is always ontological work, positioning engineers as material entities in the world amidst other entities.' So whilst neither sociological field has adequately described the co-production of objects and identities there is an emerging interest in this question. And it is one that is perhaps best understood through an onto-temporal lens.

In Chapter 4 we moved towards such an articulation, examining how engineers strategically adopted, or not, the terms 'synthetic biology' and 'synthetic biologist' alongside the shifts in the ontologies of microorganisms that had occurred as their research interests developed and their practices changed across their career timelines. We explored how early in their careers our now senior colleagues had needed to reorganise their knowledge-making practices in relation to the agency of bacteria, which began to constrain or resist what the academics wanted to do with them or know about problems on which they were working. As we have argued, bacteria, then, contributed powerfully to changes in their own ontological status over time. They were not just dancers on the stage being choreographed, but directors, too. Such periods of 'bacterial emergence', as we termed it, opened up the biological realm in new ways, and eventually led to our engineers' encounters with synthetic biology, which began new shifts in which the ontological status of bacteria would again be differently enacted.

We introduced the term 'biography' to denote the ways in which these kinds of ontological changes were narrated strategically as part of trying to enact synthetic biology. Our colleagues' own descriptions of changes in the ontological status of bacteria over their careers, and so of their own professional identities, was rhetorical, political and practical. In trying to tie themselves into the changes happening at the core of the field, whilst working to enact synthetic biology in a way that would fit the ontological clusters of the local situation, all the while trying to get jobs, keep jobs and move ahead in their jobs, they also enacted their own embodied identities in relation to bacteria through telling stories of longer-term changes. In this regard, biography is an everyday term for everyday work. It denotes the co-production of ontologies of objects and identities as well as their strategic representation within shifting practices over time. And so, like choreography, biography crucially incorporates a sense of the temporal dimensions of enactment. As such, it can sit alongside choreography and varieties of stickiness as part of a mess of concepts through which to articulate the negotiation of sameness and difference in situated ontologies.

Overall, as regards time, we have shown that governance and industrial innovation systems will shape the practices through which synthetic biology can or cannot be enacted. And, as we saw in our colleagues' success in getting water company actors to begin thinking longer-term, synthetic biology is in turn being used to shape those industrial practices. Timing and tempo in synthetic biology are part of how the field is organised at the core but are also open to change as synthetic biology is brought into different practices in new situations. Synthetic biology's emergence and novelty also have a longer history of development, not only against the background of entanglements between engineering and biology but also when situated in the everyday lives of our colleagues. Ontological shifts that can seem to have appeared quite rapidly are actually worked on in mundane ways over longer periods. The ability, or not, to bring about quick changes in practices when synthetic biology's norms began to take hold depended on longer-term efforts in which our colleagues had been engaged.

Time, both in terms of timing and tempo, is thus also enacted. They are shaped and reshaped, by emerging practices and how they mix with or change existing ones, at the level of the everyday, the local situation of academic and industrial work and at the level of discourse and governance, in the national, international and global contexts.

Of course, this is just a slice in time, as it were, for the life of the field, for the lives of our colleagues, the bacteria with which they worked, for the water industry and for us. In order to try to locate time within the project we have, to some extent, had to freeze it, by rendering it within this particular book-length account. STS scholars have to do more to experiment with how we study and represent the temporal dimensions of sociotechnical practices, particularly perhaps in an emerging field that aims to change how things are done in an ever more hasty fashion. This has important implications for how we think about the role of STS, and social science more broadly, in the context of synthetic biology.

Conclusion

Synthetic biology represents an excellent case through which to explore the emergence of a novel technoscience, the creation of new bio-objects and ontologies, the mixing of academic and industrial practices and the governance of science more broadly. We have made a case for understanding such things at the level of everyday academic life, and, in this chapter, have used the three slices of ontology, failure and time to further this investigation.

As we have progressed through the chapter we have also gestured to how these three themes are woven together. Ontologies, for example, can be difficult to enact. Practices can be jumbled up and resistant to change, making it hard to get new ontological relations to stick. Our adoption of terms around 'stickiness' to describe some of how ontologies are enacted or not in actors' efforts to change practices is one contribution to the STS literature that we hope can be adopted elsewhere.

Resistance to shifting ontologies and associated practices is one reason why there is much failure in actors' efforts to bring about change, particularly in the development of new fields. In bringing ontology and failure together, we argued that producing messy descriptions of how actors deal with failure – personally, professionally and practically – is an important part of how the ethics of ontological enactment can be conceptualised in such contexts. In this regard, it is important to attend to the affective and emotional dimensions of how change is brought about in the everyday life of academia and industry. We contribute to STS literature here as well, adding affect and emotion to Lynch's (2013) list of the practical, rhetorical and political considerations to be made in the description of ontographies.

We also looked at temporal dimensions of practice, arguing that bringing about change in whatever small measure required significant amounts of time, particularly where recalcitrant practices clustered ontologies together, resisting efforts to get new things to stick or others to detach. Resultant ontological clusters made it practically very difficult to make our academic colleagues' plans for synthetic biology run on time, at least as they were expected to when observed through the promises of the field. Rhetorically, then, there were some struggles as well. Whilst certain promises made it possible to synchronise past, present and future across academia and industry on paper and in talk, particularly through the use of grand challenges, they were harder to realise in practice. Changes in academia did not have much bearing on changes in the water industry, and our colleagues struggled to use the promises of synthetic biology to get the water company's practices to budge. The material infrastructure of the industry and the scale of such problems also meant that the tempo had to be slowed down, and expectations stretched out. Moreover, efforts to bring about change in the water industry and to work on the kinds of long-term problems faced in this arena were constrained by academic practices not immediately having to do with synthetic biology, but rather with funding regimes, reward structures and career progression. This meant that time was enacted differently in our local situation than in the core. Through constructing this argument, we have contributed to understanding time in STS by bringing the literature on promissory narratives and sociotechnical imaginaries down to the level of everyday life, and by weaving it into the emerging study of enacted ontologies.

Returning to some of the issues we raised in the introduction, we have found the use of the heuristic binary of 'core' and 'periphery' to be warranted in exploring our data and in articulating the dimensions of ontology, failure and time in synthetic biology. In our descriptions of how the core and periphery were connected, we have, in passing, talked about 'transmission', of 'carrying' and of making things 'mobile', but we have also described 'resistance', 'failure' and 'barriers' when discussing the relations between our situation and synthetic biology more broadly. This messy picture is what it was like for our actors to be working towards the use of synthetic biology in the water industry, and for us in trying to collaborate with them. However, we understand that this heuristic is not able to capture all of the mess and murkiness of how practices, objects, promises and people were entangled in our study. If others find it useful, then that is fine, but we are happy for the distinction to live and die in our project alone.

On that note, it is worth reminding readers again that our overall way of framing the book, by looking at the 'project', is also limited. Much went on that was outside of its immediate boundaries, some of which features in what we have written, but much of which does not. Moreover, such blurred lines are not merely spatial, but also temporal. We cannot extend our making sense of the project too far into either the past or the future. Nonetheless, articulating synthetic biology in this way has helped to show up some features that are hidden if one looks only at the broader picture. A fine-grained, close-up analysis of synthetic biology in situ has helped to demonstrate some of the ways in which the field is likely to remain messy, ambiguous and unstable, even as certain things close down, become standardised and stuck.

Coda 5
Reflections on Collaboration

Andy Balmer: What are the features of a good collaboration?

Academic Chemical Engineer 1: If you'd asked me six months ago I probably would have said some *Nature* paper with all our names listed at the top. But for you as a social scientist that's pretty worthless, because even though it's in *Nature* it's not just one great author, it's not sole-authored, so it's worthless to you. Talking to Susie, I think it was, I've come to the understanding that you have completely different metrics for valuing your outputs. So if we could get a spread, a range of different things that are valued by all parties, get something tangible, something useful, take something forward. We do have a proven foundation of some success in synthetic biology, so together maybe then we could write the next big three year, five year grant application, whatever it might be.

Ontologies, failure and time

In our attempts to collaborate with our colleagues in engineering and natural sciences on a synthetic biology project we conducted STS research in some traditional and some more experimental ways. We now reflect on how we helped, or not, to enact collaboration in our project, again using our three slices of ontologies, failure and time.

The 'ontological turn' (Woolgar and Lezaun, 2013) in science and technology studies represents one more reflexive loop in its development. On this occasion, taking a turn has led scholars to ponder over how STS might contribute to the ethical dimensions of ontological enactments. If there are multiple ontologies of objects enacted differently according to situated practices, then we must ask in what ways we contribute, or do not contribute, to those enactments. In this respect, there is a current concern with what kind of worlds we want to live in, and also with 'marginal' (Mol, 2012) or 'Othered' realities (Law and Singleton, 2005). The ethical spin in the ontological turn is thus to posit how STS might contribute to enacting the world in ways that are less politically violent when it comes to silencing, hiding or destroying the others of knowledge production and use.

It is important, however, to ensure that adopting technical terms such as these does not lead us away from the situated struggles of being heard, visible and supported. Looking to the everyday is one way in which to unpick some of these implications for the ethical enactment of STS research and experiments in collaboration, making use of this conceptual framework whilst retaining our grounding in empirical observations.

In our project then, we explored how barriers were constituted through situated practices, and then we tried to explain this in meetings with collaborators, for example by reflecting on how the public was being conceptualised and to what ends. We challenged, critiqued and cooperated in the analysis of barriers and in trying to overcome them. We contributed to industry days, to the organisation of the project, to meetings, to the science itself on occasion and to opening up our colleagues' understandings of how science gets done. For example, we examined the ways in which biology had been reconfigured over time, through their careers and in the recent emergence of the field, helping to reflect on the entangled practices of everyday life in academia and knowledge production. We helped to advise the iGEM team, became part of the team, helped to make a biosensor and developed some reflexive practices to overcome the sometimes difficult experiences involved in participating in academic competitions.

But we were also obstacles ourselves. For example, as regards the enactment of ontologies. To some degree, a number of our colleagues did not want to change how the barriers were enacted, just to remove them. At other times, they were trying to fix things in place, but we would not let them settle. Most evidently we got in the way of consolidating the ontology of synthetic biology. We were disruptive.

To explore this issue further, we turn to how we contributed to failure in the project, but also to success, and examine how these figured as part of our attempts to work collaboratively. As regards success, we helped to win the funding. Our contributions marked out the project application as being distinctly cross-disciplinary and that was crucial to it being successful. We helped to organise things in ways that our colleagues might not have previously considered, particularly when it came to introducing more experimental approaches to selecting and working with the iGEM team and industrial actors. We also contributed to the challenge of fitting synthetic biology into the water industry, for example in thinking about how the different spaces in which industrialists work might change how synthetic biology could be integrated. Sometimes it seemed that we were helping people to think differently, to reflect in ways that they would not have done without us being there, to ask questions and contribute our own thoughts. Of course, we also wrote papers and gave presentations. On occasion, we even lent a little cultural capital, particularly when it came to writing up the project reports and evaluations. At one point, quite a way into the project, ours were the only published outputs, and so they were useful in evidencing the work being conducted in the project to external observers. Perhaps in these and other regards we were quite valuable parts of the team in the eyes of our collaborators.

But certainly we also contributed to some of the failure. We used up time that might have been spent designing or pipetting or working on other more traditionally valued activities. We said things that were not well understood and confused people. We literally got in the way in the labs on more than one occasion. In this regard, we were definitely a part of the affective and emotional world in which knowledge was being made. This was, in fact, frequently how our presence and work became most visible. We got annoyed, and we were annoying. We got bored and we were boring. We got excited, although we were not particularly exciting. We think we were sometimes interesting and at least provocative. There were plenty of occasions when a question we would pose, or a comment we would make, would cause someone to pause, reflect, laugh, ramble, curse or sigh. We got in and we tried to collaborate. So there was a lot of failure. Partly this was because our methods involved opening up, highlighting mess and complexity.

On this note, STS interventions that demonstrate such complexity and illuminate alternative futures can be seen to threaten the success of scientific projects that are judged on their capacity to realise their initial objectives and promises. This is particularly the case in synthetic biology because of its insistence on standardisation and predictability, and of the expectation that it will deliver economic transformation in a short timeframe. As such, STS scholars, and perhaps social scientists more generally, must also negotiate our own disciplinary measures of failure and success in the context of experiments in collaboration. For example, we might consider the extent of our critical or normative engagement with the field and how this might impact on our careers and the careers of our colleagues, on the centres in which we all work, and also on the life of two academic fields, both, in differing ways, very much in the making.

This is particularly the case in the periphery, or at 'the margins' of synthetic biology, where the resources are fewer and the risks to careers, projects, departments and universities perhaps greater. What is happening at the core of synthetic biology – where the promises are cooked up, the practices designed and the money collected – has important implications for the periphery. Law and Singleton (2005: 343) argue that 'we cannot understand objects unless we also think of them as sets of present dynamics generated in, and generative of, realities that are necessarily absent.' In synthetic biology this means, for example, that work in the core depends upon work in the periphery that is erased from official accounts. Peripheral actors must work to change their practices for enacting biological objects, such as standardised parts and chassis, in order to gain from core promises and tie themselves into the changing practices, but with far less in the way of resources to do so. Without these efforts in the periphery, however, core actors' efforts to construct massive libraries of parts and to generate standards would be worthless.

There are also ways in which peripheral actors must actually pay some of the price for the success of core actors. We showed in Chapter 5, in our analysis of how failures are managed in the iGEM competition, that there are important affective dimensions to the constitution of the field that have been silenced and erased from the promissory narratives of synthetic biology. iGEM, we argued, is based not only on a promissory and moral economy, but also an affective one. This is not the case only in iGEM however. We showed that whilst working to make synthetic biology happen at the university our senior academic colleagues also had to calibrate their practices and feelings in relation to core promises, practices and affective displays.

The pursuit of standardisation, of sharing parts in registries, disembodying laboratory practices, translating work into industrial applications and of success more generally, comes at a hidden price for those in the periphery, whose responsibility it also becomes to hide the failure, the struggle, frustration and doubt in the circulation of these emerging norms. We would argue that how promissory technosciences secure their performative capacity might depend more generally upon such affective economies, in which covering up the costs of changing one's practices is rewarded with entry to the core, with the ability to claim 'sameness' and perhaps also with limited grant money and career progression. How everyday academic life and knowledge production are entangled is thus an ethical issue. Where STS has begun to open up difference and absence and how these are mutually enacted with sameness and presence, we should now ask in what situations people and objects have to work to make connections, to try to enact sameness and so to be able to access other worlds from their Othered worlds. STS has often documented the movement of knowledge and practices through the reconfiguration of networks of actors. But who pays the price for the spread of novel ontologies? The ontological turn would benefit from framing issues of marginality and Othering through the addition of affective and mundane issues such as these. The description of, and struggle to rework, onto-affective practices is at least one of the ways in which integrated STS scholars can begin to intervene, critique and collaborate in emerging fields. This would make for an important contribution to the development of post-ELSI methods for collaboration.

However, this does not mean that we should jump in, becoming para-synthetic biologists, eager to please industrial partners and make sure that everyone in the core or periphery is successful and cheery. We are not just collaborating to bring in grant funds and care for scientists or for science. Of course, there are places for disruption, for the kinds of critique with which we are familiar and in which we are well trained. Indeed, STS scholars experimenting with collaboration in synthetic biology have been very critical of certain promises being made in the field and of how these are more likely to play out in practice (Marris, 2013). In our own project, for example, we sought to make visible the kinds of changes that might come about if synthetic biology's enactment in the water industry played out according to the visions posed by our colleagues. It was obvious that R&D managers might be excited because companies might make more profits. It was less clear, however, what would become of climate change, customer's bills or of the process engineers who would potentially need more training or lose their jobs. Examining these uncertainties and inequalities in bearing the risks and costs of innovation is critical ethical work. It helps to bring in other kinds of marginality, to open up scientific and innovation practices so as to consider the effects that synthetic biology might have on those with less power, less money and less opportunity.

Who wins and who loses in the enactment of synthetic biology and its industrialisation are absolutely key questions to bringing about the field in an ethical way – a perspective and questions that were recognised, interestingly, by supposedly 'ignorant' public actors early on in its emergence (BBSRC, 2010). Questioning who wins and loses might be doable from within collaborations, though it does put those working relations at risk (Balmer *et al.*, 2015). It invites us to adapt Fujimura's (1987) work on the 'doability' of scientific problems and explore what makes for a 'doable' collaboration between social scientists, engineers and natural scientists. Where and how might we need to maintain distance in these new forms of collaborative interaction? This is important because bringing about change in the enactment of synthetic biology from within collaborations places a significant burden on social scientists, who tend to bear the responsibility for making collaborations work and for maintaining them in the face of contestation, disagreement and antagonism (Balmer *et al.*, 2015).

Are social scientists really equipped to be making science or scientists more ethical? In Chapters 2 and 5 we explored how difficult it is to shift ontologies, for example of the public or of barriers, as part of collaborative practices. We did fail to make these changes, but it was clear to us that it was not solely our failure. Experimenting with how we work together is important, but so is ensuring that our own situations do not become intolerable, and that we do not take on too much responsibility for changing practices that are as much out of our hands as they are of the engineers. Failure's corollary in ethical enactment of collaborations therefore has to be care. We have to find ways to take care of science, ourselves and 'others' as we work towards new epistemic practices.

Relatedly, and moving on to time, part of the challenge in trying to bring about changes in practices, not only for our colleagues but also for us in experimenting with collaboration, was that there was only limited time in which to do so. As we have described, there was pressure to produce outputs from the project to satisfy funders and to ensure that the group could win synthetic biology funding again in the future. However, our involvement became a problem when it came to time. We stopped engineers from doing everyday things to talk to us. We also took time to say things. It took us a long time to explain what we do with people with whom we had not worked before, and for them to get used to our forms of interaction. Some participants thought we took too long to generate 'findings' and too long to explain the nature and implications of what we had found. Much of this concern with the 'time taken' related to language use but also to our methods, which do just take time. Several of our colleagues explicitly stated that our language was opaque and that they wanted something more along the lines of a bullet-point list of quick-fix actions describing how they should act in the future to ensure the success of the field. Given the elaborate configuration of barriers we discussed in Chapter 3, it is unsurprising that participants would seek to simplify the perceived problems and to hope for quick solutions to the 'social' dilemmas they faced. It is also unsurprising that this was unachievable, even for three committed sociologists working across the various levels of the project.

In this regard, the time taken to collaborate was a problem not just because of the speed and acceleration demanded by synthetic biology's norms and promises but also because of more long-standing practices in the governance of academic life and funding regimes, of differences in methods and of power relations. On this latter point, we worked hard and took time to learn how to talk synthetic biology, but our colleagues were far less willing to take the time to learn how to talk STS. Such issues are deeply embedded into the fabric of academia and impact on the potential for collaboration across the natural and social sciences. However, this is not merely an issue of language, as tends to be the framing in the literature on interdisciplinarity. Rather, it is a situated enactment of the relations between social and natural science, performed through the practices particular to those situations and the actors concerned, and so always within different sets of power relations, temporal regimes and reward structures. Language is important, but it is just one of the everyday ways in which these broader issues manifest. To change these relations requires far more time than just learning some new terms and how to get by in a laboratory.

Time was a barrier for which our colleagues and we had not properly accounted. In the end, we ran out of time. The project ended before we had been able to make any substantive changes to 'barriers' or to overcome some of the water industry problems that our colleagues hoped to address. Synthetic biology was left in quite an ambiguous state by the end of the project. Nonetheless, new grants were funded and new relations forged. People, things, careers and ideas moved on as our collective collaboration on this project came to a close. So, for now, that leaves us with one last question.

Where are we now?

Susie maintains good working relations with her engineering colleagues although further attempts to gain external funding for their collaborative work have failed. Doctoral students continue to come and study synthetic biology using STS perspectives with her, and some of these make use of the long-standing relations that she has helped to build up within the institution. She continues to occupy the periphery of synthetic biology, collaborating, analysing, writing and drinking coffee, whilst those STSers with whom she has worked have moved on to the core.

Andy now works at another university, on a range of different projects, and with new responsibilities for teaching and management. He continues to explore the possibilities of collaboration but now as a member of one of the six RCUK-funded synthetic biology research centres (SBRCs). As has happened in recent years, social scientists in this context increasingly work under the banner of 'responsible research and innovation' (RRI) (Shapira *et al.*, 2015), in part because of how it has been embedded in the UK Synthetic Biology Roadmap (SBRCG, 2012). So, Andy now works partly within an 'RRI team'.

Kate's involvement with the field has also stuck. From beginning a PhD taking synthetic biology primarily as a case study, through which to ask broader questions about the co-production of new epistemic communities and novice practitioners, Kate now works alongside Andy in the SBRC. As such, the nature of her involvement in the field has changed. To some degree her involvement has become increasingly formalised as synthetic biology itself and the role of social scientists in relation to the field have become more institutionalised. The expectations that accompany Andy and Kate's nominal positions as 'RRI researchers' demand new kinds of participation and engagement the field, and also new kinds of resistance and care for science, self and other. But it is not yet clear how these positions might be negotiated in the long term and with what consequences for the kinds of STS work being done in these spaces (Balmer *et al.*, 2015). What exactly RRI is, how it should be practised, and if it can take a post-ELSI style, are questions still up for grabs. Nonetheless, RRI groups are more and more common, forming part of a broader trend in research governance and scientific practice in which RRI has become one focus for experiments in collaboration, though questions remain as to whether this represents business as usual or a more substantive shift (Randles *et al.*, 2014).

Looking back in 2015 on the project, partially situated from within these new formal infrastructures, our past involvement in the field takes on a particularly open, exploratory and fluid quality. What synthetic biology was, is and might become, and what our relationships with our scientific colleagues might look like, were questions very much still in their infancy at the time the project was conducted. Although our engineering collaborators would often try to close down definitions of synthetic biology and held certain expectations about what our involvement would entail, there was also a sense that we were all feeling our way through this new terrain and trying to make sense of where this might lead. This shared sense of emergence and exploration opened up space for more critical and experimental engagements with the field, for engineers and social scientists alike, and nurtured a collaborative and sometimes even a playful atmosphere. Notwithstanding the importance placed on the success of this project to the university, to departments and to individual careers, the connections and affiliations with synthetic biology at all these levels remained peripheral and ambiguous. Although we were heavily invested in the success of the project in various ways, our investment in the success of synthetic biology as an emerging discipline was less intense.

Comparably, in the new synthetic biology research centres, we have become more closely connected with the core, at least in the UK context. The significant financial and political investment in these centres means that our work, and those of the scientists and engineers involved, has become more tightly tethered to the goal of making synthetic biology an economic and political success. Although the field is still very much in an emergent phase, these centres have committed to particular contingent visions at a far larger scale and are changing and consolidating their practices in more determinate ways to try to bring those visions about accordingly. The ongoing work of changing practices in these new centres is inspiring new questions and objects of interest for sociologists. These more politically charged spaces are reconfiguring what success and failure might mean for all involved. Although there is a feeling that we might have less room to manoeuvre, there is perhaps a greater capacity here for intervening in a way that ramifies beyond our immediate locale.

Over time, the professional and personal relationships that all three of us built up with colleagues working in the field have also shifted. As people have moved on to new projects, new places and new plans, we have forged new connections. Some of the scientists and engineers on the project have returned to more traditional disciplinary arenas, whilst others are finding new niches in the interdisciplinary mix of the natural and physical sciences, distinct from synthetic biology. However, others have continued to nurture and develop their involvement with this new biotechnical field. As for the iGEM students, they all went on to be very successful in their undergraduate degrees, with the majority going on to pursue PhDs, two of whom became increasingly involved in synthetic biology at core institutions. For a summer marred by failure, they all ended up with success.

Many of our relationships with the engineers and natural scientists on the project have withered. But others have grown from strength to strength. For us, of particular importance is a supportive, collaborative community of STS scholars of which we were a part of whilst working on the project, all of whom continue to be involved in synthetic biology, albeit in different ways. Some of our STS colleagues around the UK have also become attached to the SBRCs and have become further caught up in the tide of this powerful field and now also of RRI. Others, like Susie, continue to track the emergence of synthetic biology from a more distanced position, relatively unbound from the institutional demands of these more formalised structures.

Our community of STS scholars working in synthetic biology is also now beginning to grow and change, as more social scientists are becoming integrated through the further institutional expansion of the field. The arrival of social scientists with varying histories of engagement with synthetic biology, and coming from different disciplinary backgrounds, means that a new swell of expertise has occurred, with different organisational, professional and personal interests in the field intensifying and diversifying the social scientific gaze. The burgeoning interdisciplinary mix also comes with particular challenges and opportunities, not least to confront the question of what kinds of roles the different social scientists can or will play, and the kinds of questions we can now ask as synthetic biology itself is changing apace.

To greater and lesser degrees then, we find ourselves still amidst the ever larger numbers of people, money, machines and so forth, all entangled through shifting practices, enacting differently situated forms of 'synthetic biology' into the future. In our account of experiments in synthetic biology and collaboration at the periphery, we have shown that bringing about changes is a messy, unpredictable and sometimes dangerous business. Practices might shift around, barriers might fall down, but others will resist, and stand tall. We cannot predict the future. Will the field go the way of previous promissory technosciences? What roles will we STS scholars take, what kinds of objects might we help to bring into life and what futures will we be a part of? Much in the making of synthetic biology remains unknown. What is certain, however, is that some things will change – but only some.

References

B. Adam (1994) *Time and Social Theory* (Cambridge: Policy).

M. Ashmore (1989) *The Reflexive Thesis: Wrighting Sociology of Scientific Knowledge* (Chicago, IL: University of Chicago Press).

A. S. Balmer and K. Bulpin (2013) 'Left to Their Own Devices: Post-ELSI, Ethical Equipment and the International Genetically Engineered Machine (iGEM) Competition', *BioSocieties*, 8 (3), 311–35.

A. S. Balmer, K. Bulpin, J. Calvert, M. Kearnes, A. Mackenzie, C. Marris, P. Martin, S. Molyneux-Hodgson and P. Schyfter (2012) 'Towards a Manifesto for Experimental Collaborations between Social and Natural Scientists', http://experimentalcollaborations.wordpress.com/, accessed 11 May 2013.

A. S. Balmer, J. Calvert, C. Marris, E. Frow, S. Molyneux-Hodgson, M. Kearnes, K. Bulpin, P. Schyfter, A. Mackenzie and P. Martin (2015) 'Taking Roles in Interdisciplinary Collaborations: Reflections on Working in Post-ELSI Spaces' *Science and Technology Studies* 28 (3), http://www.sciencetechnologystudies.org/v28n3.

A. S. Balmer and C. Herreman (2009) 'Craig Venter and the Re-Programming of Life: How Metaphors Shape and Perform Ethical Discourses in the Media Presentation of Synthetic Biology' in: B. Nerlich, R. Elliott and B. Larson (eds.) *Communicating Biological Sciences: Ethical and Metaphorical Dimensions* (Farnham: Ashgate).

A. S. Balmer and P. Martin (2008) *Synthetic Biology: Social and Ethical Challenges* (Biotechnology and Biological Sciences Research Council) http://www.bbsrc.ac.uk/about/policies/reviews/scientific-areas/0806-synthetic-biology/, accessed 1 February 2013.

A. S. Balmer and S. Molyneux-Hodgson (2013) 'Bacterial Cultures: Ontologies of Bacteria and Engineering Expertise at the Nexus of Synthetic Biology and Water Services', *Engineering Studies*, 5 (1), 59–73.

R. Banasiak, R. Verhoeven, R. De Sutter and S. Tait (2005) 'The Erosion Behaviour of Biologically Active Sewer Sediment Deposits: Observations from a Laboratory Study', *Water Research*, 39 (20), 5221–31.

K. Barad (2007) *Meeting the Universe Halfway: Quantum Physics and the Entanglement of Matter and Meaning* (Durham, NC: Duke University Press).

A. Barry, G. Born and G. Weszkalnys (2008) 'Logics of Interdisciplinarity', *Economy and Society*, 37 (1), 20–49.

BBSRC (2010) *Synthetic Biology Dialogue* (Biotechnology and Biological Sciences Research Council) http://www.bbsrc.ac.uk/documents/1006-synthetic-biology-dialogue-pdf/, accessed 31 July 2015.

U. Beck (1992) *Risk Society: Towards a New Modernity* (London: Sage).

S. A. Benner and A. M. Sismour (2005) 'Synthetic Biology', *Nature Reviews Genetics*, 6 (7), 533–43.

G. Bennett (2010a) *What Is a Part?* (BIOFAB Human Practices Reports) http://biofab.synberc.org/sites/default/files/HPIP_Report%201.0_v2_0.pdf, accessed 21 July 2015.

G. Bennett (2010b) *What Is iGEM?* (BIOFAB Human Practices Reports) http:// biofab.synberc.org/sites/default/files/HPIP_Report%202.0_v1.pdf, accessed 21 July 2015.

W. E. Bijker, T. P. Hughes, T. Pinch and D. G. Douglas (2012) *The Social Construction of Technological Systems: New Directions in the Sociology and History of Technology* (Cambridge, MA: MIT Press).

L. Billings and D. Endy (2008) 'Synthetic Biology: Cribsheet No. 16.', *SEED Magazine*, http://seedmagazine.com/images/uploads/16cribsheet.pdf, accessed 23 July 2015.

K. Birch and D. Tyfield (2013) 'Theorizing the Bioeconomy Biovalue, Biocapital, Bioeconomics Or ... What?', *Science, Technology and Human Values*, 38 (3), 299–327.

A. Biro (2007) 'Water Politics and the Construction of Scale', *Studies in Political Economy*, 80, 9–30.

D. Bloor (1976) *Knowledge and Social Imagery* (London: Routledge and Kegan Paul).

S. Blue (2012) 'Ongoing Change in the Rhythms of Mixed Martial Arts Practice', *International Journal of Sport and Society*, 3 (3), 161–70.

M. Borup, N. Brown, K. Konrad and H. Van Lente (2006) 'The Sociology of Expectations in Science and Technology', *Technology Analysis and Strategic Management*, 18 (3–4), 285–98.

G. C. Bowker and S. L. Star (2000) *Sorting Things Out: Classification and Its Consequences* (Cambridge, MA: MIT Press).

C. Brown (2004) 'Biobricks to Help Reverse-Engineer Life', *EE Times*, http://www. eetimes.com/document.asp?doc_id=1150423, accessed 05 July 2015.

N. Brown (2003) 'Hope against Hype: Accountability in Biopasts, Presents and Futures', *Science Studies*, 16 (2), 3–21.

N. Brown and A. Kraft (2006) 'Blood Ties: Banking the Stem Cell Promise', *Technology Analysis and Strategic Management*, 18 (3–4), 313–27.

N. Brown and M. Michael (2003) 'A Sociology of Expectations: Retrospecting Prospects and Prospecting Retrospects', *Technology Analysis and Strategic Management*, 15 (1), 3–18.

N. Brown, B. Rappert and A. Webster (2000) *Contested Futures: A Sociology of Prospective Techno-Science* (Farnham: Ashgate).

L. L. Bucciarelli (1994) *Designing Engineers* (Cambridge, MA: MIT Press).

K. Bulpin and S. Molyneux-Hodgson (2013) 'The Disciplining of Scientific Communities', *Interdisciplinary Science Reviews*, 38 (2), 91–105.

C. Cagnin, E. Amanatidou and M. Keenan (2012) 'Orienting European Innovation Systems Towards Grand Challenges and the Roles That FTA Can Play', *Science and Public Policy*, 39 (2), 140–52.

J. Calvert (2008) 'The Commodification of Emergence: Systems Biology, Synthetic Biology and Intellectual Property', *BioSocieties*, 3 (4), 383–98.

J. Calvert (2010) 'Synthetic Biology: Constructing Nature?', *The Sociological Review*, 58, 95–112.

J. Calvert (2012) 'Ownership and Sharing in Synthetic Biology: A "Diverse Ecology"' of the Open and the Proprietary?', *BioSocieties*, 7 (2), 169–87.

J. Calvert (2013) 'Collaboration as a Research Method? Navigating Social Scientific Involvement in Synthetic Biology' in: N. Doorn, D. Schuurbiers, I. V. D. Poel and M. E. Gorman (eds.) *Early Engagement and New Technologies: Opening up the Laboratory* (Netherlands: Springer).

J. Calvert and P. Martin (2009) 'The Role of Social Scientists in Synthetic Biology', *EMBO Reports,* 10 (3), 201–04.

A. Cambrosio and P. Keating (1988) '"Going Monoclonal": Art, Science, and Magic in the Day-to-Day Use of Hybridoma Technology', *Social Problems,* 35 (3), 244–60.

L. Campos (2010) 'That Was the Synthetic Biology That Was', in M. Schmidt, A. Kelle, A. Ganguli-Mitra, and H. Vriend (eds.) *Synthetic Biology: The Technoscience and Its Societal Consequences* (Netherlands: Springer).

L. Campos (2012) 'The Biobrick™ Road', *BioSocieties,* 7 (2), 115–39.

R. Carlson (2007) 'Laying the Foundations for a Bio-Economy', *Systems and Synthetic Biology,* 1 (3), 109–17.

R. Carlson (2010) *Biology Is Technology: The Promise, Peril and New Business of Engineering Life* (Cambridge, MA: Harvard University Press).

M. Cave (2009) *Independent Review of Competition and Innovation in Water Markets* (Department for Environment, Food and Rural Affairs) www.Defra.Gov.Uk/Environment/Water/Industry/Cavereview, accessed 20 July 2015.

G. Church (2001) *Next Generation Techniques* (Speech delivered at the SB5.0 conference in Stanford, California) https://vimeo.com/26905527, accessed 19 September 2015.

L. Clarke (2013) *Synthetic Biology – The UK Roadmap* (Speech delivered at the Multidisciplinary Synthetic Biology Research Centres Information Workshop hosted by RCUK) http://www.bbsrc.ac.uk/documents/1306-sbrc-information-workshop-pdf/, accessed 24 July 2015.

C. Cockerton (2011) *Going Synthetic: How Scientists and Engineers Imagine and Build a New Biology.* Ph.D. Thesis (London: LSE Ph.D. Thesis Repository).

A. Coghlan (2012) 'Biology Is a Manufacturing Capability', *New Scientist,* https://www.newscientist.com/article/mg21628946-100-biology-is-a-manufacturing-capability/, accessed 20 July 2015.

H. M. Collins (1974) 'The TEA Set: Tacit Knowledge and Scientific Networks', *Science Studies,* 4 (2), 165–85.

H. M. Collins (1988) 'Public Experiments and Displays of Virtuosity: The Core-Set Revisited', *Social Studies of Science,* 18 (4), 725–48.

H. M. Collins (1992 [1985]) *Changing Order: Replication and Induction in Scientific Practice,* 2nd Edition (Chicago: University of Chicago Press).

H. M. Collins (2010) *Gravity's Shadow: The Search for Gravitational Waves* (Chicago, IL: University of Chicago Press).

N. Crossley (2001) *The Social Body: Habit, Identity and Desire* (London: Sage).

C. Cussins (1996) 'Ontological Choreography: Agency through Objectification in Infertility Clinics', *Social Studies of Science,* 26 (3), 575–610.

DBIS (2013) *Over £60 Million for Synthetic Biology* (Department for Business Innovation and Skills Press Release) https://www.gov.uk/government/news/over-60-million-for-synthetic-biology, accessed 20 July 2015.

M. De Laet and A. Mol (2000) 'The Zimbabwe Bush Pump Mechanics of a Fluid Technology', *Social Studies of Science,* 30 (2), 225–63.

P. Dear (2001) 'Science Studies as Epistemography' in: J. A. Labinger and H. Collins (eds.) *The One Culture* (Chicago, IL: University of Chicago Press).

S. Delamont and P. Atkinson (2001) 'Doctoring Uncertainty: Mastering Craft Knowledge', *Social Studies of Science,* 31 (1), 87–107.

A. Deplazes-Zemp (2012) 'The Conception of Life in Synthetic Biology', *Science and Engineering Ethics*, 18 (4), 757–74.

M. Douglas (1966) *Purity and Danger: An Analysis of Concepts of Pollution and Taboo* (London: Routledge).

G. L. Downey and J. C. Lucena (2004) 'Knowledge and Professional Identity in Engineering: Code – Switching and the Metrics of Progress', *History and Technology*, 20 (4), 393–420.

ETC (2009) *The Story of Synthia* (Action Group on Erosion, Technology and Concentration) http://www.etcgroup.org/content/story-synthia, accessed 20 July 2015.

ETC (2010) *Synthia Is Alive...and Breeding: Panacea or Pandora's Box?* (Action Group on Erosion, Technology and Concentration) http://www.etcgroup.org/content/synthia-alive-%E2%80%A6-and-breeding-panacea-or-pandoras-box, accessed 20 July 2015.

D. Endy (2005) 'Foundations for Engineering Biology', *Nature*, 438 (7067), 449–53.

D. Endy (2011) *Message from the President* (SB5.0 Delegate Program Book) http://sb5.biobricks.org/files/sb5-program-book-v3.pdf, accessed 31 July 2015.

L. Eriksson (2012) 'Pluripotent Promises: Configurations of a Bio-Object' in: N. Vermeulen, S. Tamminen and A. Webster (eds.) *Bio-Objects: Life in the 21st Century.* (Burlington, VT: Ashgate).

U. Felt (2015) 'Keeping Technologies Out: Sociotechnical Imaginaries and the Formation of Austria's Technopolitical Identity', in S. Jasanoff and S-H. Kim (eds.) *Dreamscapes of Modernity: Sociotechnical Imaginaries and the Fabrication of Power* (Chicago: Chicago University Press).

S. C. Finlay (2013) 'Engineering Biology? Exploring Rhetoric, Practice, Constraints and Collaborations within a Synthetic Biology Research Centre', *Engineering Studies*, 5 (1), 26–41.

E. Fisher, M. O'Rourke, R. Evans, E. B. Kennedy, M. E. Gorman and T. P. Seager (2015) 'Mapping the Integrative Field: Taking Stock of Socio-Technical Collaborations', *Journal of Responsible Innovation*, 2 (1), 39–61.

D. Fitzgerald and F. Callard (2015) 'Social Science and Neuroscience Beyond Interdisciplinarity: Experimental Entanglements', *Theory, Culture and Society*, 32 (1), 3–32.

D. Fitzgerald, N. Jones, S. Choudhury, M. Friedner, N. Levin, S. Lloyd, T. Meyers, N. Myers and E. Raikhel (2014a) *The Collaborative Turn: Interdisciplinarity across the Human Sciences* (Somatosphere.Net) http://somatosphere.net/2014/07/the-collaborative-turn-interdisciplinarity-across-the-human-sciences.html, accessed 14 July 2015.

D. Fitzgerald, M. M. Littlefield, K. J. Knudsen, J. Tonks and M. J. Dietz (2014b) 'Ambivalence, Equivocation and the Politics of Experimental Knowledge: A Transdisciplinary Neuroscience Encounter', *Social Studies of Science*, 44 (5), 701–21.

S. Franklin (2008) 'Debate: Beyond the Genome: The Challenge of Synthetic Biology', *BioSocieties*, 3, 3–20.

E. Frow and J. Calvert (2013a) '"Can Simple Biological Systems Be Built from Standardized Interchangeable Parts?" Negotiating Biology and Engineering in a Synthetic Biology Competition', *Engineering Studies*, 5 (1), 42–58.

E. Frow and J. Calvert (2013b) 'Opening up the Future(s) of Synthetic Biology', *Futures*, 48, 32–43.

E. Frow (2013) 'Making Big Promises Come True: Articulating and Realizing Value in Synthetic Biology', *BioSocieties*, 8 (4), 432–48.

J. H. Fujimura (1987) 'Constructing 'Do-Able' Problems in Cancer Research: Articulating Alignment', *Social Studies of Science*, 17 (2), 257–93.

J. H. Fujimura (1988) 'The Molecular Biological Bandwagon in Cancer Research: Where Social Worlds Meet', *Social Problems*, 35 (3), 261–83.

J. H. Fujimura (1996) *Crafting Science: A Sociohistory of the Quest for the Genetics of Cancer* (Boston, MA: Harvard University Press).

L. Garforth and A. Cervinková (2009) 'Times and Trajectories in Academic Knowledge Production' in: U. Felt (ed.) *Knowing and Living in Academic Research* (Prague: Institute of Sociology of the Academy of Sciences of the Czech Republic).

GeneArt (2015) *Gene Synthesis Technologies* (Life Technologies) http://www. lifetechnologies.com/uk/en/home/life-science/cloning/gene-synthesis/gene-art-gene-synthesis.html, accessed 21 July 2015.

D. G. Gibson, J. I. Glass, C. Lartigue, V. N. Noskov, R.-Y. Chuang, M. A. Algire, G. A. Benders, M. G. Montague, L. Ma and M. M. Moodie (2010) 'Creation of a Bacterial Cell Controlled by a Chemically Synthesized Genome', *Science*, 329 (5987), 52–56.

T. F. Gieryn (1999) *Cultural Boundaries of Science: Credibility on the Line* (Chicago, IL: University of Chicago Press).

Ginkgo (2015a) *About Ginkgo Bioworks* (Ginkgo Bioworks) http://ginkgobioworks. com/about/, accessed 11 March 2015.

Ginkgo (2015b) *About the Foundry* (Ginkgo Bioworks) http://ginkgobioworks. com/foundry/, accessed 11 March 2015.

A. D. Ginsberg, J. Calvert, P. Schyfter, A. Elfick and D. Endy (2014) *Synthetic Aesthetics: Investigating Synthetic Biology's Designs on Nature* (Boston, MA: MIT Press).

C. Goodwin (1995) 'Seeing in depth', *Social Studies of Science*, 25 (2), 237–274.

D. H. Guston and D. Sarewitz (2002) 'Real-Time Technology Assessment', *Technology in Society*, 24 (1–2), 93–109.

S. Güttinger (2013) 'Creating Parts That Allow for Rational Design: Synthetic Biology and the Problem of Context-Sensitivity', *Studies in History and Philosophy of Biological and Biomedical Sciences*, 44 (2), 199–207.

J. M. Halberstam and I. Livingston (1995) *Posthuman Bodies* (Bloomington, IN: Indiana University Press).

P. Hancock, B. Hughes, E. Jagger, K. Paterson, R. Russell, E. Tulle-Winton and M. Tyler (2000) *The Body, Culture and Society* (Buckingham: Open University Press).

J. Hannigan (2006) *Environmental Sociology*, 2nd Edition (London: Routledge).

D. J. Haraway (1991) *Simians, Cyborgs, and Women: The Reinvention of Nature* (London: Routledge).

HCSTC (2013) *Bridging the Valley of Death: Improving the Commercialisation of Research: Eighth Report of the Session 2012–13, HC 348* (House of Commons Science and Technology Committee) http://www.publications.parliament.uk/, accessed 27 July 2015.

A. Hedgecoe (2003) 'Terminology and the Construction of Scientific Disciplines: The Case of Pharmacogenomics', *Science, Technology and Human Values*, 28 (4), 513–37.

A. Hedgecoe and P. Martin (2003) 'The Drugs Don't Work Expectations and the Shaping of Pharmacogenetics', *Social Studies of Science*, 33 (3), 327–64.

I. Hellsten and B. Nerlich (2011) 'Synthetic Biology: Building the Language for a New Science Brick by Metaphorical Brick', *New Genetics and Society*, 30 (4), 375–97.

J. Henkel and S. M. Maurer (2009) 'Parts, Property and Sharing', *Nature Biotechnology*, 27 (12), 1095–98.

S. Hilgartner (2015) 'Capturing the Imaginary: Vanguards, Visions and the Synthetic Biology Revolution' in: S. Hilgartner, C. Miller and R. Hagendijk (eds.) *Science and Democracy: Making Knowledge and Making Power in the Biosciences and Beyond* (London: Routledge).

S. Holm and R. Powell (2013) 'Organism, Machine, Artifact: The Conceptual and Normative Challenges of Synthetic Biology', *Studies in History and Philosophy of Science Part C: Studies in History and Philosophy of Biological and Biomedical Sciences*, 44 (4PtB), 627–31.

T. Holmberg, N. Schwennesen and A. Webster (2011) 'Bio-Objects and the Bio-Objectification Process', *Croatian Medical Journal*, 52 (6), 740–42.

B. Hughes (2000) 'Medicalized Bodies' in: P. Hancock, B. Hughes, E. Jagger, K. Paterson, R. Russell, E. Tulle-Winton and M. Tyler (eds.) *The Body, Culture and Society* (Buckingham: Open University Press).

JCVI (2010) *First Self-Replicating Synthetic Cell* (J. Craig Venter Institute) http://www.jcvi.org/cms/research/projects/first-self-replicating-synthetic-bacterial-cell/overview/, accessed 20 July 2015.

S. Jasanoff (2004) 'The Idiom of Co-Production' in: S. Jasanoff (ed.) *States of Knowledge: The Co-Production of Science and Social Order* (New York, NY: Routledge).

S. Jasanoff and S-H. Kim (2013) 'Sociotechnical Imaginaries and National Energy Policies', *Science as Culture*, 22 (2), 189–96.

K. Jordan and M. Lynch (1992) 'The Sociology of a Genetic Engineering Technique: Ritual and Rationality in the Performance of the Plasmid Prep' in: A. Clarke and J. Fujimura (eds.) *The Right Tools for the Job: At Work in Twentieth-Century Life Sciences* (Princeton, NJ: Princeton University Press).

K. Jordan and M. Lynch (1998) 'The Dissemination, Standardization and Routinization of a Molecular Biological Technique', *Social Studies of Science*, 28 (5–6), 773–800.

M. Kaiser, M. Kurath, S. Maasen and C. Rehmann-Sutter (2010) *Governing Future Technologies: Nanotechnology and the Rise of an Assessment Regime*, Sociology of the Sciences Yearbook Volume 27 (Netherlands: Springer).

M. Kearnes (2013) 'Performing Synthetic Worlds: Situating the Bioeconomy', *Science and Public Policy*, 40 (4), 453–65.

M. Kearnes and M. Weinroth (2011) *A New Mandate? Research Policy in a Technological Society* (Durham: University of Durham Research Reports).

A. Kelle (2009) 'Synthetic Biology and Biosecurity', *EMBO Reports*, 10 (1S), S23–S27.

E. F. Keller (2009) 'What Does Synthetic Biology Have to Do with Biology?', *BioSocieties*, 4 (2–3), 291–302.

E. Kerr and L. Garforth (2015) 'Affective Practices, Care and Bioscience: A Study of Two Laboratories', *The Sociological Review*, DOI: 10.1111/1467-954X.12310.

R. Kitney and P. Freemont (2012) 'Synthetic Biology – the State of Play', *FEBS letters*, 586 (15), 2029–36.

T. Knight (2003) *Idempotent Vector Design for Standard Assembly of Biobricks* (MIT Synthetic Biology Working Group) http://web.mit.edu/synbio/release/docs/biobricks.pdf, accessed 20 July 2015.

K. D. Knorr Cetina (1982) 'Scientific Communities or Transepistemic Arenas of Research? A Critique of Quasi-Economic Models of Science', *Social Studies of Science*, 12 (1), 101–30.

K. D. Knorr Cetina (1999) *Epistemic Cultures: How the Sciences Make Knowledge* (Cambridge, MA: Harvard University Press).

R. Kwok (2010) 'Five Hard Truths for Synthetic Biology', *Nature*, 463 (7279), 288–90.

M. Lampland and S. L. Star (2009) *Standards and Their Stories: How Quantifying, Classifying, and Formalizing Practices Shape Everyday Life* (Ithaca, NY: Cornell University Press).

B. Latour (1987) *Science in Action: How to Follow Scientists and Engineers through Society* (Boston, MA: Harvard University Press).

B. Latour (1992) 'One More Turn after the Social Turn: Easing Science Studies into the Non-Modern World' in: E. Mccullin (ed.) *The Science Studies Reader* (Notre Dame, IN: University of Notre Dame Press).

B. Latour (1996) *Aramis, or, the Love of Technology* (Cambridge, MA: Harvard University Press).

B. Latour and S. Woolgar (1986) *Laboratory Life: The Construction of Scientific Facts* (Princeton, NJ: Princeton University Press).

R. Lave, P. Mirowski and S. Randalls (2010) 'Introduction: STS and Neoliberal Science', *Social Studies of Science*, 40 (5), 659–75.

J. Law (2002) 'On Hidden Heterogeneities: Complexity, Formalism and Aircraft Design' in: J. Law and A. Mol (eds.) *Complexities: Social Studies of Knowledge Practices* (London: Duke University Press).

J. Law and M. Lien (2012) 'Slippery: Field Notes on Empirical Ontology', *Social Studies of Science*, 43 (3), 363–78.

J. Law and V. Singleton (2005) 'Object Lessons', *Organization*, 12 (3), 331–55.

T. Lenoir (1997) *Instituting Science: The Cultural Production of Scientific Disciplines* (Stanford, CA: Stanford University Press).

D. Lupton (1997) 'Foucault and the Medicalisation Critique' in: A. Petersen and R. Bunton (eds.) *Foucault, Health and Medicine* (London: Routledge).

M. Lynch (2001) 'The Epistemology of Epistopics: Science and Technology Studies as an Emergent (Non)Discipline', *American Sociological Association: Science, Knowledge and Technology Section (ASA-SKAT) Newsletter*, Fall, 2–3.

M. Lynch (2013) 'Postscript Ontography: Investigating the Production of Things, Deflating Ontology', *Social Studies of Science*, 43 (3), 444–62.

N. Machiavelli (1961) *The Prince*. Translated with an Introduction by George Bull (London: Penguin Books).

A. Mackenzie (2010) 'Design in Synthetic Biology', *BioSocieties*, 5 (2), 180–98.

A. Mackenzie (2013) 'The economic principles of industrial synthetic biology: cosmogony, metabolism and commodities', *Engineering Studies*, 5 (1), 74–89.

D. MacKenzie (2006) 'Is Economics Performative? Option Theory and the Construction of Derivatives Markets', *Journal of the History of Economic Thought*, 28 (1), 29–55.

D. MacKenzie (2008) *An Engine, Not a Camera: How Financial Models Shape Markets* (Boston, MA: MIT Press).

D. MacKenzie and G. Spinardi (1995) 'Tacit Knowledge, Weapons Design, and the Uninvention of Nuclear Weapons', *American Journal of Sociology*, 101 (1), 44–99.

C. Marris (2013) 'Synthetic Biology's Malaria Promises Could Backfire', *SciDevNet*, http://www.scidev.net/global/biotechnology/opinion/synthetic-biology-s-malaria-promises-could-backfire.html, accessed 24 July 2015.

C. Marris (2015) 'The Construction of Imaginaries of the Public as a Threat to Synthetic Biology', *Science as Culture*, 24 (1), 83–98.

C. Marris, C. Jefferson and F. Lentzos (2014) 'Negotiating the Dynamics of Uncomfortable Knowledge: The Case of Dual Use and Synthetic Biology', *BioSocieties*, 9 (4), 393–420.

P. Martin, N. Brown and A. Turner (2008) 'Capitalizing Hope: The Commercial Development of Umbilical Cord Blood Stem Cell Banking', *New Genetics and Society*, 27 (2), 127–43.

R. McDaniel and R. Weiss (2005) 'Advances in Synthetic Biology: On the Path from Prototypes to Applications', *Current Opinion in Biotechnology*, 16 (4), 476–83.

I. Metzler and A. Webster (2011) 'Bio-Objects and Their Boundaries: Governing Matters at the Intersection of Society, Politics, and Science', *Croatian Medical Journal*, 52 (5), 648–50.

M. Meyer (2010) 'The Rise of the Knowledge Broker', *Science Communication*, 32 (1), 118–27.

M. Meyer and S. Molyneux-Hodgson (2010) 'The Dynamics of Epistemic Communities', *Sociological Research Online*, 15 (2), 14. 10.5153/sro.2154.

C. C. Mody (2005) 'The Sounds of Science: Listening to Laboratory Practice', *Science, Technology and Human Values*, 30 (2), 175–98.

A. Mol (2002) *The Body Multiple: Ontology in Medical Practice* (London: Duke University Press).

A. Mol (2012) 'Mind Your Plate! The Ontonorms of Dutch Dieting', *Social Studies of Science*, 43 (3), 379–96.

S. Molyneux-Hodgson and A. S. Balmer (2014) 'Synthetic Biology, Water Industry and the Performance of an Innovation Barrier', *Science and Public Policy*, 41 (4), 507–19.

S. Molyneux-Hodgson and M. Meyer (2009) 'Tales of Emergence – Synthetic Biology as a Scientific Community in the Making', *BioSocieties*, 4 (2), 129–45.

S. Molyneux-Hodgson and J.W.N. Smith (2007) 'Building a research agenda on water policy: an exploration of the Water Framework Directive as an interdisciplinary problem', *Interdisciplinary Science Reviews*, 32, 187–202.

N. Moran (2007) 'Public Sector Seeks to Bridge "Valley of Death"', *Nature Biotechnology*, 25 (3), 266.

N. Myers and J. Dumit (2011) 'Haptic Creativity and the Mid-Embodiments of Experimental Life' in: F. Mascia-Lees (ed.) *A Companion to the Anthropology of the Body and Embodiment*. (Chichester: Blackwell Publishing).

P. Nightingale and P. Martin (2004) 'The Myth of the Biotech Revolution', *Trends in Biotechnology*, 22 (11), 564–69.

A. Nordmann and A. Rip (2009) 'Mind the Gap Revisited', *Nature Nanotechnology*, 4 (5), 273–74.

A. Nordmann and A. Schwarz (2010) 'Lure of the "Yes": The Seductive Power of Technoscience' in: M. Kaiser, K. Kurath, S. Maasen and C. Rehmann-Sutter (eds.) *Governing Future Technologies: Nanotechnology and the Rise of an Assessment Regime* (Netherlands: Springer).

M. A. O'Malley (2009) 'Making Knowledge in Synthetic Biology: Design Meets Kludge', *Biological Theory*, 4 (4), 378–89.

M. A. O'Malley, A. Powell, J. F. Davies and J. Calvert (2008) 'Knowledge-Making Distinctions in Synthetic Biology', *BioEssays*, 30 (1), 57–65.

Ofwat (2015) *Information on Markets* (Ofwat) http://www.ofwat.gov.uk/competition/, accessed 6 March 2015.

G. Osborne (2012) *Speech by the Chancellor of the Exchequer, Rt Hon George Osborne MP, to the Royal Society*, https://www.gov.uk/government/speeches/speech-by-the-chancellor-of-the-exchequer-rt-hon-george-osborne-mp-to-the-royal-society, accessed 24 July 2015.

R. Owen, J. Stilgoe, P. Macnaghten, M. Gorman, E. Fisher and D. Guston (2013) 'A Framework for Responsible Innovation' in: R.Owen, J. Bessant and M. Heintz (eds.) *Responsible Innovation: Managing the Responsible Emergence of Science and Innovation in Society* (Chichester: John Wiley and Sons Ltd.).

E. Parens, J. Johnston and J. Moses (2009) *Ethical Issues in Synthetic Biology: An Overview of the Debates* (Woodrow Wilson International Center for Scholars) http://www.synbioproject.org/process/assets/files/6334/synbio3.pdf, accessed 27 July 2015.

M. Pickersgill (2011) 'Connecting Neuroscience and Law: Anticipatory Discourse and the Role of Sociotechnical Imaginaries', *New Genetics and Society*, 30 (1), 27–40.

T. J. Pinch and H. M. Collins (1984) 'Private Science and Public Knowledge: The Committee for the Scientific Investigation of the Claims of the Paranormal and Its Use of the Literature', *Social Studies of Science*, 14 (4), 521–46.

M. Polanyi (1967) *The Tacit Dimension* (Chicago, IL: Chicago University Press).

A. Powell, M. A. O'Malley, S. Müller-Wille, J. Calvert and J. Dupré (2007) 'Disciplinary Baptisms: A Comparison of the Naming Stories of Genetics, Molecular Biology, Genomics, and Systems Biology', *History and Philosophy of the Life Sciences*, 29 (1), 5–32.

P. Rabinow and G. Bennett (2007) *From Bio-Ethics to Human Practice* (Anthropology of the Contemporary Research Collaboratory Working Papers), http://anthropos-lab.net/wp/publications/2007/08/workingpaperno11.pdf, accessed 27 July 2012.

P. Rabinow and G. Bennett (2009) 'Synthetic Biology: Ethical Ramifications 2009', *Systems and Synthetic Biolology*, 3 (1–4), 99–108.

P. Rabinow and G. Bennett (2012) *Designing Human Practices: An Experiment with Synthetic Biology* (Chicago, IL: University of Chicago Press).

RAE (2009) *Synthetic Biology: Scope, Applications and Implications* (Royal Academy of Engineering) http://www.raeng.org.uk/publications/reports/synthetic-biology-report, accessed 20 March 2015.

S. Randles, B.R. Dorbeck-Jung, R. Lindner, and A. Rip (2014) 'Where to Next for Responsible Innovation?' in: C. Coenen, A. Dijkstra, C. Fautz, J. Guivant, C. Milburn, and H. van Lente (eds.), *Innovation and Responsibility: Engaging with New and Emerging Technologies* (Amsterdam: IOS Press).

J. Rappaport (2008) 'Beyond Participant Observation: Collaborative Ethnography as Theoretical Innovation', *Collaborative Anthropologies*, 1 (1), 1–31.

A. Rip, T. Misa and J. Schot (1995) *Managing Technology in Society: The Approach of Constructive Technology Assessment* (London: Thomson).

P. Robbins (2009) *The Genesis of Synthetic Biology: Innovation, Interdisciplinarity and the iGEM Student Competition* (American Sociological Association Annual Meeting) http://oro.open.ac.uk/19052/2/D6E8892E.pdf, accessed 21 July 2015.

M. C. Roco and W. S. Bainbridge (2002) 'Converging Technologies for Improving Human Performance: Integrating from the Nanoscale', *Journal of Nanoparticle Research*, 4 (4), 281–95.

N. Rose (2007a) 'Molecular Biopolitics, Somatic Ethics and the Spirit of Biocapital', *Social Theory and Health*, 5 (1), 3–29.

N. Rose (2007b) *The Politics of Life Itself: Biomedicine, Power, and Subjectivity in the Twenty-First Century* (Princeton, NJ: Princeton University Press).

N. Rose and C. Novas (2005) 'Biological Citizenship' in: A. Ong and S. J. Collier (eds.) *Global Assemblages: Technology, Politics, and Ethics as Anthropological Problems* (Oxford: Blackwell Publishing).

D. Savage, J. Way and P. Silver (2008) 'Defossiling Fuel: How Synthetic Biology Can Transform Biofuel Production', *ACS Chemical Biology*, 3 (1), 13–16.

J. Sawicki (1991) *Disciplining Foucault: Feminism, Power, and the Body* (London: Routledge).

SBRCG (2012) *A Synthetic Biology Roadmap for the UK* (Synthetic Biology Roadmap Coordination Group) http://www.rcuk.ac.uk/RCUK-prod/assets/documents/publications/SyntheticBiologyRoadmap.pdf, date accessed 20 July 2015.

M. Schmidt (2008) 'Diffusion of Synthetic Biology: A Challenge to Biosafety', *Systems and Synthetic Biology*, 2 (1–2), 1–6.

M. Schmidt, A. Ganguli-Mitra, H. Torgersen, A. Kelle, A. Deplazes and N. Biller-Andorno (2009) 'A Priority Paper for the Societal and Ethical Aspects of Synthetic Biology', *Systems and Synthetic Biology*, 3 (1–4), 3–7.

P. Schyfter (2012) 'Technological Biology? Things and Kinds in Synthetic Biology', *Biology and Philosophy*, 27 (1), 29–48.

P. Schyfter (2013a) 'How a 'Drive to Make' Shapes Synthetic Biology', *Studies in History and Philosophy of Science Part C: Studies in History and Philosophy of Biological and Biomedical Sciences*, 44 (4), 632–40.

P. Schyfter (2013b) 'Propellers and Promoters: Emerging Engineering Knowledge in Aeronautics and Synthetic Biology', *Engineering Studies*, 5 (1), 6–25.

P. Schyfter, E. Frow and J. Calvert (2013) 'Guest Editorial: Synthetic Biology: Making Biology into an Engineering Discipline', *Engineering Studies*, 5 (1), 1–5.

S. Shapin (1989) 'The Invisible Technician', *American Scientist*, 77 (6), 554–63.

P. Shapira and A. Gök (2015) '*UK Synthetic Biology Centres tasked with addressing public concerns*', *The Guardian*, http://www.theguardian.com/science/political-science/2015/jan/30/uk-synthetic-biology-centres-tasked-with-addressing-public-concerns, accessed 20 September 2015.

P. Shapira, J. Youtie and Y. Li (2015) 'Social Science Contributions Compared in Synthetic Biology and Nanotechnology', *Journal of Responsible Innovation*, 2 (1), 143–148.

R. P. Shetty (2008) *Applying Engineering Principles to the Design and Construction of Transcriptional Devices*. Ph.D. Thesis. (Cambridge, MA: Massachusetts Institute of Technology).

R. P. Shetty, D. Endy and T. F. Knight Jr (2008) 'Engineering Biobrick Vectors from Biobrick Parts', *Journal of Biological Engineering*, 2 (1), 1–12.

E. Singer (2009) 'Startup That Builds Biological Parts: Ginkgo Bioworks Aims to Push Synthetic Biology to the Factory Level', *MIT Technology Review*, http://www.technologyreview.com/news/415546/startup-that-builds-biological-parts/, accessed 11 March 2015.

S. Slaughter and G. Rhoades (1996) 'The Emergence of a Competitiveness Research and Development Policy Coalition and the Commercialization of

Academic Science and Technology', *Science, Technology and Human Values,* 21 (3), 303–39.

C. Smith (2013a) *Discovery in Synthetic Biology a Step Closer to New Industrial Revolution* (Imperial College London Press Release) http://www3.imperial. ac.uk/newsandeventspggrp/imperialcollege/newssummary/news_31-1-2013– 12–18–1, accessed 24 July 2015.

C. Smith (2013b) *New Centre to Harvest Economic Benefits of Synthetic Biology* (Imperial College London Press Release) http://www3.imperial.ac.uk/news-andeventspggrp/imperialcollege/newssummary/news_10–7-2013–12–14–56, accessed 27 July 2015.

S. Star and J. Griesemer (1989) 'Institutional Ecology,'Translations' and Boundary Objects: Amateurs and Professionals in Berkeley's Museum of Vertebrate Zoology, 1907–39', *Social Studies of Science,* 19 (3), 387–420.

S. L. Star (2010) 'This Is Not a Boundary Object: Reflections on the Origin of a Concept', *Science, Technology and Human Values,* 35 (5), 601–17.

A. Stirling (2005) 'Opening up or Closing Down? Analysis, Participation and Power in the Social Appraisal of Technology', *Science, Technology and Human Values,* 33 (2), 262–94.

A. Swidler (2001) 'What Anchors Cultural Practices' in: T. R. Schatzki, K. K. Cetina and E. V. Savigny (eds.) *The Practice Turn in Contemporary Theory* (London: Routledge).

T. Swierstra, M. Boenink, B. Walhout and R. Van Est (2009) 'Converging Technologies, Shifting Boundaries', *Nanoethics,* 3 (3), 213–16.

T. Swierstra and A. Rip (2007) 'Nano-Ethics as Nest-Ethics: Patterns of Moral Argumentation About New and Emerging Science and Technology', *Nanoethics,* 1 (1), 3–20.

E. Swyngedouw (1997) 'Neither Global nor Local:'Glocalization'and the Politics of Scale' *Spaces of Globalization: Reasserting the Power of the Local* (London: Longman).

E. Swyngedouw (2005) 'Dispossessing H_2O: The Contested Terrain of Water Privatization', *Capitalism Nature Socialism,* 16 (1), 81–98.

J. Tait (2010) 'Governing Synthetic Biology: Processes and Outcomes' in: M. Schmidt, A. Kelle, A. Ganguli-Mitra, H. de Vriend (eds.) *Synthetic Biology: The Technoscience and Its Societal Consequences* (Netherlands: Springer).

J. Tait (2012) 'Adaptive Governance of Synthetic Biology', *EMBO Reports,* 13 (7), 579.

Thames Water (2015) *The Water Cycle* (Thames Water Utilities Limited) http:// www.thameswater.co.uk/cycles/, accessed 21 July 2015.

S. Timmermans and M. Berg (1997) 'Standardization in Action: Achieving Local Universality through Medical Protocols', *Social Studies of Science,* 27 (2), 273–305.

S. Traweek (1988) *Beamtimes and Lifetimes: The World of High Energy Physicists* (Cambridge, MA: Harvard University Press).

B. S. Turner (1992) *Regulating Bodies: Essays in Medical Sociology* (London: Routledge).

B. S. Turner (2008) *The Body and Society: Explorations in Social Theory,* 3rd Edition (London: Sage).

United Utilities (2015) 'Preparing Water' http://www.unitedutilities.com/ all-about-water.aspx, accessed 19 September 2015.

C. Venter (2005) *Sampling the Ocean's DNA* (TED Global 2005) http://www.ted.com/talks/craig_venter_on_dna_and_the_sea/transcript?language=en, accessed 31 July 2015.

J. C. Venter and D. G. Gibson (2010) 'How We Created the First Synthetic Cell', *The Wall Street Journal*, http://www.wsj.com/articles/SB10001424052748704026204575266460432676840, accessed 28 July 2015.

N. Vermeulen (2012) 'Growing a Cell in Silico: On How the Creation of a Bio-Object Transforms the Organisation of Science' in: N. Vermeulen, S. Tamminen and A. Webster (eds.) *Bio-Objects: Life in the 21st Century* (Farnham: Ashgate).

N. Vermeulen, S. Tamminen and A. Webster (2012) *Bio-Objects: Life in the 21st Century* (Farnham: Ashgate).

W. G. Vincenti (1990) *What Engineers Know and How They Know It* (Baltimore: Johns Hopkins University Press).

D. Vinck (2003) *Everyday Engineering: An Ethnography of Design and Innovation* (Cambridge, MA: MIT Press).

C. Waldby (2002) 'Stem Cells, Tissue Cultures and the Production of Biovalue', *Health: an Interdisciplinary Journal for the Social Study of Health, Illness and Medicine*, 6 (3), 305–23.

P. Walde (2010) 'Building Artificial Cells and Protocell Models: Experimental Approaches with Lipid Vesicles', *BioEssays*, 32 (4), 296–303.

C. Waterton (2002) 'From Field to Fantasy Classifying Nature, Constructing Europe', *Social Studies of Science*, 32 (2), 177–204.

E. Wenger (1999) *Communities of Practice: Learning, Meaning, and Identity* (Cambridge: Cambridge University Press).

WFD (2000) *Water Framework Directive* (European Parliament) http://eur-lex.europa.eu/legal-content/EN/TXT/?uri=CELEX:32000L0060, accessed 20 July 2015.

D. Willetts (2013) *Eight Great Technologies* (London: Policy Exchange).

J. Wilsdon and R. Willis (2004) *See-through Science: Why Public Engagement Needs to Move Upstream* (Demos) http://www.demos.co.uk/publications/paddlingupstream, accessed 27 July 2012.

S. Woolgar and J. Lezaun (2013) 'The Wrong Bin Bag: A Turn to Ontology in Science and Technology Studies?', *Social Studies of Science*, 43 (3), 321–40.

S. Woolgar and J. Lezaun (2015) 'Missing the (Question) Mark? What Is a Turn to Ontology?', *Social Studies of Science*, 45 (3), 462–67.

WWICS (2010) *Trends in Synthetic Biology Research Funding in the United States and Europe* (Woodrow Wilson International Center for Scholars) http://www.wilsoncenter.org/sites/default/files/final_synbio_funding_web2.pdf, accessed 20 July 2015.

S. Wyatt (2007) 'Making Time and Taking Time, a Review of "Time Innovation and Mobilities" by Peter Peters', *Social Studies of Science*, 37 (5), 821–24.

L. Yarris (2004) *Synthetic Biology Offers New Hope for Malaria Victims* (Berkeley Lab Press Release) http://www2.lbl.gov/Science-Articles/Archive/sb-PBD-anti-malarial.html, accessed 20 July 2015.

M. Ylönen and L. Pellizzoni (2012) *Neoliberalism and Technoscience: Critical Assessments* (Farnham: Ashgate).

Yorkshire Water (2015) *How We Treat Your Water* (Yorkshire Water Utilities Limited) http://www.yorkshirewater.com/your-water-services/drinking-water/supplying-water/how-we-treat-your-water.aspx, accessed 21 July 2015.

Index

Printed in Great Britain
By Bookmasters